T0325461

Pharmaceutical Emulsions

Pharmaceutical Emulsions

A Drug Developer's Toolbag

Dipak K. Sarker

School of Pharmacy and Biomolecular Sciences
University of Brighton
UK

WILEY Blackwell

This edition first published 2013 © 2013 by John Wiley & Sons, Ltd

Wiley Blackwell is an imprint of John Wiley & Sons, formed by the merger of Wiley's global Scientific, Technical and Medical business with Blackwell Publishing.

Registered office: John Wiley & Sons, Ltd, The Atrium, Southern Gate, Chichester, West Sussex, PO19 8SQ, UK

Editorial offices: 9600 Garsington Road, Oxford, OX4 2DQ, UK
   The Atrium, Southern Gate, Chichester, West Sussex, PO19 8SQ, UK
   111 River Street, Hoboken, NJ 07030-5774, USA

For details of our global editorial offices, for customer services and for information about how to apply for permission to reuse the copyright material in this book please see our website at www.wiley.com/wiley-blackwell.

Library of Congress Cataloging-in-Publication Data

Sarker, Dipak K., author.
 Pharmaceutical emulsions : a drug developer's toolbag / by Dipak K. Sarker.
   p. ; cm.
 Includes bibliographical references and index.
 ISBN 978-0-470-97683-8 (cloth)
 I. Title.
 [DNLM: 1. Emulsions–chemistry. 2. Biopharmaceutics–standards. 3. Chemistry, Pharmaceutical–methods. QV 786.5.C7]
 RS420
 615.1′9–dc23
                                                                    2013018795

A catalogue record for this book is available from the British Library.

Wiley also publishes its books in a variety of electronic formats. Some content that appears in print may not be available in electronic books.

Set in 10.5/13pt Times by Laserwords Private Limited, Chennai, India.
Printed and bound in Singapore by Markono Print Media Pte Ltd

1   2013

Dedicated to my two fabulous sons
Hugh Callum Sarker
and
Noah Marcus Vesco Sarker

Contents

Mathematical symbols (with normal units)

Greek

σ stress (Pa)

τ shear stress (Pa), decay time (s^{-1}), delay time (s^{-1})

γ_{dot} shear rate (Hz, s^{-1})

Γ surface excess concentration (mol/m^2)

κ Debye length (m)

μ micron (10^{-6} m), ionic strength (mol/dm^3, M)

η bulk viscosity, coefficient of viscosity (mPa s, cP, cSt)

η^* complex bulk viscosity (Pas)

η_d surface (dilational) viscosity (mN s/m)

η_s surface (shear) viscosity (N/s/m)

ε dielectric constant (dimensionless), permittivity ($8.854 \times 10^{-10}\,C^2/N/m^2$)

ε strain ($\Delta L/L$, where L is length)

ζ zeta potential (mV)

ρ density, bulk phase (g/cm^3)

$\Delta\rho$ density difference (medium minus dispersed phase) (g/cm^3)

γ surface tension (mN/m)

ϕ phase volume (dimensionless)

ψ surface potential (mV)

θ contact angle ($^\circ$)

π surface pressure (mN/m)

λ wavelength (nm)

N   frequency, wavenumber (cm)

Latin

H Hamaker constant for constants based on interaction of oil and water $[(\text{oil}-\text{oil})^{0.5} - (\text{water}-\text{water})^{0.5}]^2$ (1×10^{-20} J)

N_a Avogadro's number (6.022×10^{23})

R gas constant (8.314 J/K/mol)

R_g radius of gyration (m)

g acceleration due to gravity (9.81 m/s^2)

G	Young's modulus (Pa)
$G*$	complex modulus (Pa)
G'	storage modulus (Pa)
G''	loss modulus (Pa)
ΔP	Laplace pressure (Pa, N/m^2)
K	dissociation constant of ionisation
P	partition coefficient ([octanol]/[water])
k_B	Boltzmann constant (1.381×10^{-23} J/K)
k	rate constant (s^{-1})
T	temperature ($^\circ$C, K)
t	time (s)
E_d	dilational modulus (mN/m)
E	complex elastic modulus (mN/m)
$D_{4,3}$	usual hydrodynamic particle diameter (by volume) (nm)
$D_{3,2}$	Sauter hydrodynamic particle diameter (by area) (nm)
D	diffusion coefficient (cm^2/s)
exp or e	exponent, exponential
	Other symbols are explained at their point of use.

Acronyms and abbreviations

A/W	air–water system
ADME	adsorption, distribution, metabolism and excretion (of a drug)
AFM	atomic force microscopy
CMC	critical micelle concentration
DDS	drug delivery system
DLS	dynamic light scattering
DLVO	a theory from Derjaguin, Landau, Verwey and Overbeek
DSC	differential scanning calorimetry
EMEA	European Medicines Agency (a.k.a. EMA)
EPR	enhanced permeability and retention (effect)
FDA	Food and Drug Administration (USA)
FDG	2-fluoro-2-deoxy-d-glucose
GCP	good clinical practice
GI tract	gastrointestinal tract
GLP	good laboratory practice
GMP	good manufacturing practice
HLB	hydrophile–lipophile balance
ICH	International Conference on Harmonisation
ISO	International Standards Organization (Geneva)
IV	intravenous
LCST/HCST	lower or higher critical solution temperature
LNC	lipid nanocapsule
LNP	lipid nanoparticle
MHRA	Medicines Health and Regulatory (products) Agency (UK)
NPD	new (medicinal) product development
O/W	oil-in-water system
PCL	poly(caprolactone)
PCS	photon correlation spectroscopy
PEG	poly(ethylene glycol)
PIT	phase-inversion temperature
PLA	poly(lactic acid)
PLGA	poly(lactic–glycolic acid)
PMMA	polymethyl(methacrylate)
QA	quality assurance
QC	quality control
QMS	quality management system

RES/MPS	reticuloendothelial system/monocyte–phagocyte system
SC	stratum corneum (skin)
SEM	scanning electron microscopy
SLC	solid lipid capsule
SLN	solid lipid nanoparticle
TD	transdermal
TLF	thin liquid film
TQMS	total quality management system
W/O	water-in-oil
W/O/W	water-in-oil-in-water
WHO	World Health Organization
	Other abbreviations are explained at their point of use.

Preface

The primary target audience for this text is students engaged in MPharm and professional practice modules, pharmacy technician or internal formulation and NPD courses and MScs in industrial pharmacy, PGDip industrial pharmaceutical studies and related themes. Most schools of pharmacy have about 40–160 of these students per cohort. Yet the book must simultaneously be pertinent to MRess, MScs, PhDs and postdocs (and BScs) in 'pharmaceutical' sciences. It is thus specialist yet generalist, to the point of not simply being a collection of research papers, but instead a teaching/training text. In light of this, the book is not targeted exclusively at either the undergraduate, the advanced researcher or the experienced industrialist. It will be seen by industrial pharmacists (pharmaceutical scientists, chemists, engineers, etc) as generalist, and for real subject experts it merely represents a referential 'pocket guide' and not an encyclopaedic reference manual. Unlike many colloid and dispersions books, this text is not generalist in the sense of application universality and it is exclusively written for those involved with *pharmaceutical* emulsions (a 'hot' topic, to quote one reviewer). In this sense, it has absolute value and novelty in terms of being rather specific. Many other books are available which elaborate theory and physics or physical chemistry background (e.g. Adamson, 1990; Hiemenz and Rajagopalan, 1997; Goodwin, 2000 and more recent editions). In principle, this book is primarily targeted at pharmacists, pharmaceuticists, medics and pharmacologists, and its form alludes to this in a significant manner.

I started my involvement with 'colloids' (now 'nanotechnology' in current 'in-speak') as an undergraduate dealing with industrial dispersions, then as a masters' chemical engineering student dealing with fabrication. During a physics PhD and numerous postdocs I had the pleasure to work with and in research groups in the UK, France, Germany and Italy, where dispersions (foams, thin liquid films and emulsions) were the mainstay of the target product or the vehicle for mechanistic elucidation. Dispersions investigated included model food foams and emulsions, liquid ion-exchange systems, theoretical and mechanistic models and industrial products of a food, automotive, petrochemical and medicinal nature. I have had the great luck to have worked and collaborated with some truly great thinkers and international colloid celebrities: Peter Wilde, David Clark, Jim Mingins, Vic Morris, Brian Robinson, Eric Dickinson, Monique Axelos, Yves Popineau, Daniel Bonn, Jacques Meunier, Vance Bergeron, Zdravko Lalchev, Reinhard Miller, Jürgen Krägel, Clive Washington, Seyed Moghimi, Vladimir Torchilin and Sandy Florence, to name but a few. Today, and for the last decade or more, most of my interest has been in pharmaceutical dispersions. Coarse emulsions (hereafter simply referred to as 'emulsions'), nanoemulsions, micelles – simple, reverse and swollen – are the building blocks and centre points of nanomedicine, the pharmaceutical and therapeutic environment and modelling of drug encapsulation, new product design (nanoparticle drug delivery systems), increased efficacy and dosage miniaturisation, interfacial sculpting and molecular nanoengineering.

My research expertise, on which this book is founded, traverses areas of biophysics, material sciences, pharmaceutics and biopharmaceutics, food science, chemical engineering, physical chemistry, rheology and polymer science, medicinal chemistry, chemical biology, engineering, industrial product design and regulation and analytical chemistry. For teaching, I use a wide variety of books or chapters, and there are a number of really good pharmaceutics textbooks, but their main failings for pharmacy students is that they traverse year one to year four basic concepts such as pK_a and log D and feature only one chapter (generalist) dealing with the pivotal role of 'emulsions' in medical products (therapeutics) sciences. I hope to expand on this information without providing a cost-restrictive or excessively detailed text and to focus entirely upon the dispersed particle or particle within matrix technologies, which is perhaps better suited for students with some basic primary experience or knowledge of pharmacy dispersions. Using many figures and tables is the chosen, I have attempted to provide a summary of the salient facts and thus keep the text short. As with all things, there is a compromise to be made between what we know and would like to say and the restrictions of time and the funds available in the student's pocket. I hope students and professionals alike will find the book useful, suitably informative and yet portable and readable.

Dipak K. Sarker
Brighton, 2013

Acknowledgements

Thanks to Ralitza, Hugh, Noah, Brenda, Anita, Dilip and Asis and the other members of family Sarker. I am what I am because of you. I have done what I have done with the aid of you.

'To See a World in a Grain of Sand
And a Heaven in a Wild Flower,
Hold Infinity in the Palm of Your Hand
And Eternity in an Hour.'

'Augeries of Innocence': William Blake (1757–1827)

About the companion website

This book is accompanied by a companion website:

www.wiley.com/go/sarker/pharmaceuticalemulsions

The website includes:

- Further case studies
- Powerpoints of all figures from the book for downloading
- PDFs of all tables from the book for downloading

I
Product considerations: medicinal formulations

All medicines and therapeutics formulated as 'emulsions' must typically and characteristically contain consistent amounts and specified polymorphic forms of the relevant drug (Mahato, 2007). Depending on the route of administration, some intervention may be required in order to pasteurise or sterilise the sample and thus reduce the risk of microbial growth and pathogenicity or the production of toxins. Unfortunately, most pharmaceutical dispersions, including emulsions, are thermodynamically unstable (metastable) and undergo significant change upon heating or irradiation as a consequence of changes in bulk and interfacial rheology (Sarker *et al.*, 1999).

A patient receiving a cytotoxic drug intravenously in the form of a fine emulsion dispersion requires as a minimum that the product be suitably pure (P), suitably consistent (C) between batches and free from injurious elements, for example pathogenic microorganisms, spores and toxins, and thus of appropriate quality (Q). Acceptable PCQ is essential in any high-quality medicine (Sarker, 2008). Topical medicines (Sarker, 2006b) do not necessarily require the same degree of sterility (unless applied to broken skin), but it is worth striving for, as poor quality (Di Mattia *et al.*, 2010) lessens product shelf life.

Shape, size, polydispersity and surface coverage (Sarker *et al.*, 1999) all impact on the shelf life and efficacy (Sarker, 2005a, 2006b) of a drug delivery system (DDS). The most significant difficulty, industrially speaking, is the routine manufacture of a product and its profile, given variabilities and extremes of thermal and mechanical processing and resultant changes in surface chemistry and composition (Sarker *et al.*, 1999; Sarker, 2005b).

An emulsion (derived from the Latin *mulgeo* and/or Ancient Greek $\alpha\mu\varepsilon\lambda\gamma\omega$ (*amelgo*): terms for milk, which is an emulsion) is a mixture of unmixable, immiscible or in principle unblendable fractions. It implies a discontinuity and heterogeneity on a microscopic (nanoscopic) scale. In classic terms, both phases are usually liquid, but this can be challenged in a plethora of common forms. The two phases can be entwined in the form of a crude 'amalgam', as in a depot (as with many liquid phases) of transdermal patches, of a colloidal particle

Pharmaceutical Emulsions: A Drug Developer's Toolbag, First Edition. Dipak K. Sarker.
© 2013 John Wiley & Sons, Ltd. Published 2013 by John Wiley & Sons, Ltd.

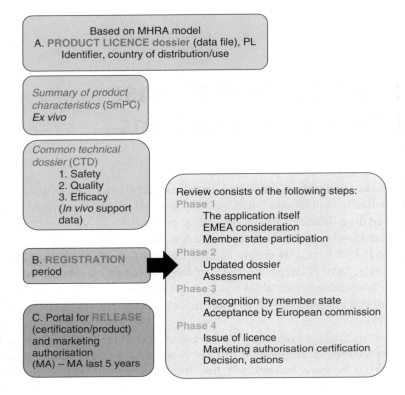

Figure I.1   Route for marketing authorisation (MA; product licence) of commercial pharmaceutical products. The schematic uses the MHRA as a template. Other regulators have a varied but generically similar approach. The figure is divided into (a) the licence itself, (b) registration and (c) the portal for release

(e.g. micelle), as in a microemulsion, or microheterogeneously, as in the case of a fine emulsion (cream, lotion, ointment). However, it is also the case that both the interior of a normal micelle and the bilayer leaflet of a vesicle represent a phase in which water is immiscible but apolar conjugated, aromatic or lipophilic molecules are miscible. The dispersion (see Section 2.1.1 and 2.3.1) is aided by inclusion of an emulsifier (emulgent, surfactant, etc.), which helps mixing and dispersion (Becher, 2001).

In order to make scholarly and industrial use of emulsions, the right level of background in physical chemistry and the theory of emulsion formation and stability are important, providing understanding of emulsion behaviour and the construction of effective DDSs. Critical considerations for medicinal formulations are: safety, efficacy, dose, hygiene, consistency, purity, quality, reproducibility, toxicity, effects, impurities and extraneous matter, cost, legal compliance, supplier reputation (for the fabrication) and the finished form.

All new emulsion products must pass through a regulatory process as just outlined. Figure I.1 shows a summary of the UK Medicines and Healthcare products Regulatory Agency (MHRA) regulation process for the approval of a new dispersion drug dosage form. Other regulators, such as the US Food and Drug Administration (FDA), also demand presentation of satisfactory safety and efficacy data (the presentation form may, however, change).

A drug product receives approval based on a product licence (PL or marketing authorisation, MA) dossier. Here the summary of product characteristics (SmPC) includes, for example, descriptions of the drug, the routes of manufacture and the physicochemical characteristics (*ex vivo*) of the molecule, batch information and stability testing information. In humans, the new chemical entity (NCE, candidate molecule) becomes known as an investigational new drug (IND), as presented in the common technical dossier (CTD). The absolute content of the CTD changes between different global regulators, but all CTDs contain information on the human response to the molecule and are responsible for chronicling the clinical evaluation in three tiers of increasingly complex (scrutiny, complexity, modelling, testing, severity and reliability) studies. The central aim of the clinical trials is to prove the product's quality (purity and consistency), based on toxicology and other *in vivo* tests. This then permits the IND to pass on to a new drug (product) application (NDA) and terminate development, proceeding to product registration and subsequent release.

The PL (or MA) is customarily granted for five years (Sarker, 2008).

1

Historical perspective

Emulsions in various guises have been around since the dawn of time
(e.g. mammalian milk, opal gemstones). What we describe as an 'emulsion' is
today a very measured and well-understood entity (Becher, 2001), as a result
of a chronology of profound and insightful scientific discoveries (see Table 1.1)
and industrial practices (Valtcheva-Sarker *et al.*, 2007; Sarker, 2010). Some very
'big' names feature in the list of events behind 'emulsion' and associated colloid
(nanotechnology) science (Gregoriadis, 1973, 1977; Sarker *et al.*, 1999; Pashley
and Karaman, 2004; Sarker, 2009a,b, 2012a).

1.1 Landmarks

These are largely definable by their subsequent impact on the area of colloid science and pharmaceutics (Florence and Attwood, 1998). There have been some
noteworthy exceptions to good practice, which have impacted on key considerations in later drug development:

- *1937* Sulphanilamide elixir, containing diethylene glycol, kills 107; established a need for drug safety before marketing.

- *1958* The US Food and Drug Administration (FDA) publishes in the Federal
  Register the first list of substances 'generally recognized as safe' (GRAS).

This paved the way to formalisation of the regulation of drugs:

- *1958* Thalidomide (Kevadon) licensed for use in the UK.

- *1961* McBride writes about increased frequency of malformations
  (phocomeliae).

- *1962* Amendments to FDA-US legislature are made based on findings
  (Sarker, 2008).

Pharmaceutical Emulsions: A Drug Developer's Toolbag, First Edition. Dipak K. Sarker.
© 2013 John Wiley & Sons, Ltd. Published 2013 by John Wiley & Sons, Ltd.

Table 1.1 Historical landmarks in the development of the fundamental and applied sciences relevant to the manufacture and use of 'pharmaceutical emulsions'

Fundamental science	Discovery	Applied (e.g. pharmaceutical) science	Discovery
1661, Hooke	Capillarity	1904, Pickering	Solid particle-stabilised emulsions
1805, Young and Laplace	Curvature equations, wetting	1909, Erhlich	Targeted delivery (magic bullet)
1827, Brown	Particulate motion	1932, Langmuir	Surface adsorption of amphiphiles
1860, Graham	Existence of colloids	1961, Bangham	Liposome
1907, Ostwald	Notion of disperse/ continuous phase	1975, L'Oréal	Invention of the niosome
ca. 1905, Einstein	Viscosity, shape and frictional models	ca. 1993, Gasco-Müller-Lucks *et al.*	Invention of the solid lipid nanoparticle (SLN)
ca. 1910, Gouy and Chapman	Description of electrical double layer		
1913, McBain	Idea of the micelle		
1941–1948, Derjaguin/Landau & Verwey/Overbeek	Theory of colloid stability		

Numerous historical 'colloid science' figures have been omitted.

Many products of the type covered here are initially microheterogeneous dispersions, so any means of enhancing or creating 'better' uniformity is most welcome to industrialists and clinicians alike (Sarker, 2008; Benson and Watkinson, 2012). Revolutionary improvements in general, fundamental and pharmaceutical understanding are indicated in Table 1.1. The history of colloid and emulsion science is profound and covers more than 5 millennia of invention and design. Current products which employ emulsions and various 'forms' of emulsification (e.g. entrapment of drug in the liposome leaflet) have uses which include flavour encapsulation, multiple emulsions for drug encapsulation and microemulsions or multiple emulsions (Florence and Attwood, 1998; Sarker, 2008).

It is a gross simplification to present the list above without recognising the diverse scientific applications that have arisen from additional theories, too numerous to mention in this book, but which include estimation of dispersion forces (London, ca. 1920), double-layer theories (Gouy and Chapman, ca. 1910) and pioneering work on microemulsions (Schulman, ca. 1959).

1.2 Significant discoveries

Pickering emulsions (Ramsden, 1903; Pickering, 1907; Binks, 2002; Aveyard *et al.*, 2003; Arditty *et al.*, 2004; Concannon *et al.*, 2010) are not a new concept but are gaining in interest in pharmacy applications. They are based on solid rather than molecular emulsifier coverage of dispersed droplets, are able to promote far 'greater product stability' than dispersed droplets and have been used as drug delivery platforms in a two-tier drug delivery system (DDS) (Concannon *et al.*, 2010). Nanoemulsion products of this type can be used to improve therapeutic efficacy (Sarker, 2005a; Valtcheva-Sarker *et al.*, 2007). Lipinski (Lipinski *et al.*, 1997; Lipinski, 2000) has described a series of rules for and established an understanding of the mechanism by which drug molecules traverse cell membranes and deliver a 'payload' of drug to the cell of interest (Valtcheva-Sarker *et al.*, 2007), and these have been pivotal to the understanding of the rigours of preformulation and good product manufacture. Solid lipid nanoparticle (SLN) DDSs and related *de novo* technologies and products were first discussed by a well-established group of researchers (Eldem *et al.*, 1991; Gasco, 1993; Müller *et al.*, 1995; Müller and Lucks, 1996) in the 1990s. These types of gel-phase lipid delivery system offer huge potential due to their control over drug entrapment, as seen with the novel drug product Qutenza (capsaicin).

The liposome, since its initial conception in 1959, has developed into many forms, some of which are used routinely to produce long circulating nanoparticles of use, for example, in cancer treatment (see Section 2.1.2 and Chapter 7). Liposome technology products and related chemicals such as stealth liposomes (using poly(ethylene glycol), PEG), niosomes, chitosomes, polymersomes, virosomes and so on, which may also be used diagnostically (Maurer *et al.*, 2001; Tiwari *et al.*, 2012), usually act as a form of lipid dispersion for delicate and lipophilic drugs and have revolutionised drug delivery (see Section 2.1.2 and Chapter 7). Liposomes and micelles can 'emulsify/solubilise' lipophilic drugs in the aliphatic portion of their superstructure (Sarker, 2010). Thermodynamically stable microemulsions, proposed by Shulman in 1959, provide an alternative means of solubilisation for apolar drugs. These types of dispersed system are often also referred to as 'swollen micelles', 'transparent emulsions' and 'solubilised oil'. New forms of emulsion, such as solid (SLN), core and shell lipid nanoparticles (LNCs, SLCs, LNPs), along with polymer micelles (which may or may not include lipid derivatisation) now look very promising in terms of extensions to a multitude of products, such as biomimetic and biocompatible materials and composite materials (Sarker 2006a, 2010, 2012a; Tiwari *et al.*, 2012). Such biomimetic materials are only likely to grow in number and applications to medicine (e.g. state-of-the-art treatments, radioimaging/radiopharmacy, etc.).

1.3 Difficulties

The pharmaceuticist is faced with problems over the route of delivery, possible destabilisation (phase inversion (PIT), Ostwald ripening, creaming, cracking and flocculation) and chemical changes (such as autoxidation and rancidity, and a subsequent modification of efficacy based on form and release profile). These are impacted on greatly by:

- The product's form (formulation aids, excipients) and its interrelationships (Sarker, 2002; Di Mattia *et al.*, 2010) and intrinsic stability.

- How the product is made.

- How and where in the body the product is intended to be used (Figure 1.1).

- How successfully and categorically the product can be tested and controlled in terms of consistency.

Fluidity of the low-molecular-weight (LMW) emulsifier adsorbed layer is necessary for surface repair (Sarker *et al.*, 1995a,b). This is not so important when solids or polymers are used to stabilise the oil/water interface. Increasing either temperature or oil phase volume and emulsifier (surfactant) concentration can cause phase inversion, which can be devastating for parenteral emulsions, for example (Araujo

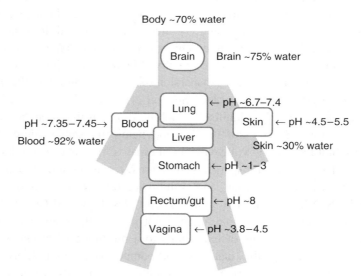

Figure 1.1 Cartoon of human anatomy/physiology from a drug delivery perspective, including principle routes of drug delivery (and obstacles) and the constitutions or environmental variations that impact on the (choice of) drug and the selection of a vehicle

et al., 2007; D'Ascenzo *et al.*, 2011). Natural fats are also susceptible to autoox-idation (Sarker, 2005a, 2012a), which can result in a modification of phase or form, leading to a reduction in the shelf life and in the chemical stability of the emulsion and any encapsulated drug. Key considerations are:

- Release and control of release.

- Hygienic status and sterility/pasteurisation, and their impact on effective drug delivery.

Effective drug delivery is based upon three notions:

- Efficient encapsulation of the drug.

- Successful 'targeting' of the drug to a 'specific' region of the body.

- Successful release of the drug *in situ*.

These difficulties are further compounded by the location to which the drug product is delivered and by variations in pH, ionic strength, temperature and permeability (see Figure 1.1).

Any discrepancy in the compatibility of excipients and formulation and the activity of the encapsulated drug (Sarker, 2004a,b) can lead to poor product perfor-mance (see Section 15.1). For any kind of emulsion or emulsifier/lipid dispersion, a key consideration is the temperature changes experienced by the product. Temper-ature influences solubility and the solubility of a drug in lipid obeys Bancroft's rule (oil in which the surfactant is more soluble will tend to form the dispersed phase of an emulsion), which can lead to a catastrophic phase inversion. This can be overcome in part by using several emulsifiers and employing hydrophile–lipophile balance (HLB) matching (see Table 1.1) of surfactant mixtures to best formulate the size and type of the emulsion droplets. The product is also defined by its potency, which covers the drug content. We can define entrapment efficiency for any drug thus:

$$\text{entrapment efficiency } (\%) = (\text{amount entrapped/total amount}) \times 100 \quad (1.1)$$

This entrapment can be augmented by using solubilising aids and penetration enhancers, such as PEG. The expression does not indicate where the drug might be located in the particle or its homogeneous dispersion.

Although emulsions (coarse type) are inherently unstable, given that there are different formulation types and different excipients, we are able to discuss 'relative stability' or 'product stability' by comparing emulsions to a standard or to one another. Thus, product stability is important in accounting for the behaviour of emulsions under the different conditions they encounter (e.g. in the gastrointestinal

(GI) tract (gut pH and absorption efficacy; see Section 12.4.3)), and the effect this has on emulsion stability and drug release (see Figure 1.1). The mechanisms for the release of the drug from the oil phase are critical.

Other issues that deserve appropriate consideration include: product syneresis (weeping), Ostwald ripening, crystallisation, recrystallisation, sequestration, coacervation, dose and drug flux estimation, particle size/texture/rheology/coverage and uniformity. Flocculation of the product and subsequent creaming and 'cracking' are driven by Derjaguin–Landau–Verwey–Overbeek (DLVO) colloidal stability theory (see Section 3.2.1). It is also important to make sure the manufacturing environment is clean and organised before initiation of the production run. This reduces the chances of making mistakes and of wasteful manufacture. It is mandatory before commencement of any pharmaceutical production that the manufacturing conditions do not compromise the content, form or efficacy of the product.

1.4 Traditional uses

The two most common manifestations of 'simple' emulsions (Figure 1.2) are found in topical and injectable medications. These involve:

- Creams (oil and water in approximately equal proportions) and ointments that combine oil and water (composition varies from 80 : 20 to 20 : 80) or lotions (mostly water).
- Parenteral and intravenous (IV) preparations.

In more recent times, variation combination products have been routinely seen. These include:

- Gels and pastes (three agents: oil, water and solid).

Lesser products for pharmacy use include:

- Parenteral nutrition products.
- Vitaminised nutraceutical suspensions and vitamin supplements.

These products are usually fabricated by combinatorial mixing of emulsifiers of varying HLB (apolarity), shape and form, as discussed at length in Section 2.3 (see Table 1.2, Figure 1.3). This gives the product designer the scope for greater emulsifier interfacial complexation (Sarker *et al.*, 1995a) and better product stability.

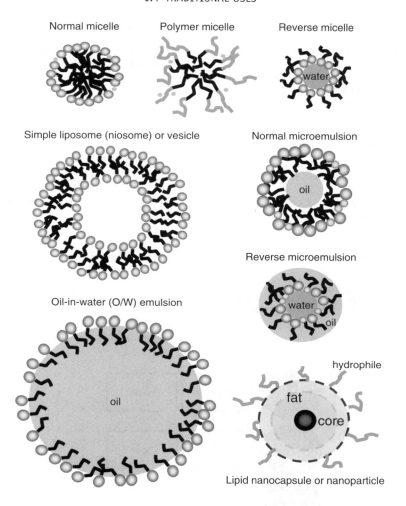

Figure 1.2 Phenomenological representation of the basic forms of 'emulsions' and 'emulsified products' used in pharmacy. The schematic shows a two-dimensional representation (not to scale) of three-dimensional entities. Emulsifier hydrophile–lipophile balance (HLB) and moiety geometry drive the form created

The complexity, scope and variety of dispersions, making use of fats, oils, lipids and emulsifiers, are represented schematically in Figure 1.2. The emulsion, liposome and micelle represent the most widely used nanoparticles in contemporary pharmacy (Sarker, 2005a, 2006a). Traditionally, the vast majority of emulsions have been used for internal application (parenteral drugs) and externally for topical (skin and mucous membrane) use. However, in recent

Table 1.2 Hydrophile-lipophile balance (HLB) or ratio of some notable emulsifiers (surfactants) used in pharmacy

Sample	Value (20 °C)	Description
Sodium dodecyl sulphate	40	Ionic
Hexadecyl trimethyl ammonium bromide	18	Ionic (disinfectant *ex vivo*)
Tween 20	16.7	Polyoxyethylene-based
PEG400 monooleate	13.1	Derivatised polymer
Methylcellulose	11	Polymer
Gelatin	9.8	Protein
Span 20	8.6	Sorbitan fatty acid ester
Span 80	4.3	Sorbitan fatty acid ester
Lecithin, e.g. soya	4	Amphoteric emulsifier
Glyceryl monostearate	3.8	Lecithin derivative
Span 85	1.8	Sorbitan fatty acid ester
Oleic acid	1	Fatty acid

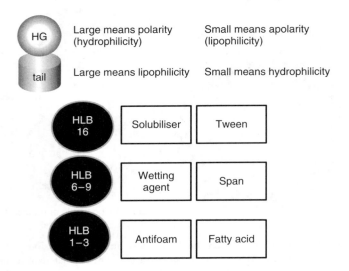

Figure 1.3 Hydrophile–lipophile balance (HLB) and surfactant–emulsifier shape, and its impact on use

approaches they have been used more widely as parts of transdermal patch products (Benson and Watkinson, 2012).

HLB mixing is used in creating nanostructured vehicles. Novel emulsifier blends and emulsifiers (e.g. Janus particles), along with fabrication of structured surfaces to withstand process-induced destabilisation, may also be used to compensate for small compositional variances. A new wave of pharmaceutics research (Bivas-Benita *et al.*, 2004; Sarker, 2010, 2012a) is considering the use of emulsions in pulmonary (Bivas-Benita *et al.*, 2004), nasal (Kumar *et al.*, 2008), colonic and tablet (Corveleyn and Remon, 1998; Pouton, 2000; Hansen *et al.*, 2005) routes to drug administration.

1.5 Product regulation

Regulation of pharmaceutical emulsions follows the same criteria as that of other pharmaceutical products in general (Sarker, 2008). Where it differs is in the inclusion of potent or noxious actives such as radioisotopes, e.g. Indium (^{111}In), or cytotoxic (anticancer, etoposide, cisplatin, doxorubicin) or even controlled drugs such as diamorphine. This regulation is centred on the risks of hypo- or hyper-dosing of the drug (Roberts, 1981). Many of the safety concerns are eradicated at the preclinical stage of development, through the validation of a consistently formulated product (Sarker, 2008). During oversight by the relevant drug regulatory bodies (FDA, International Conference on Harmonisation of Technical Requirements for Registration of Pharmaceuticals for Human Use (ICH), World Health Organization (WHO)), concerns over consistent fabrication should be addressed and aligned with regular quality control intervention to secure product acceptability, known as PCQ (purity, consistency, quality (i.e. safety and efficacy); Sarker, 2008). In pharmacy, excipients are usually restricted to approved GRAS-grade materials. A system of practice can reduce the risk of noncompliance. Such systems are driven by the regulatory bodies themselves and by the industrial manufacturer. They include adherence to specifications and guidelines and adoption of a total quality management system (GMP, QA, QC) in order to ensure the quality of the product (see Figure I.1).

2

What is an emulsion?

An emulsion is a biphasic, metastable coarse dispersion of two immiscible materials, usually liquids (typically oil and water), that produces a semisolid (Sarker, 2008). A spherical droplet (dispersed phase) is distributed in a dispersion medium (or continuous phase). This can occur for more than a few seconds only if the droplet is covered (Figure 2.1) with a layer of emulsifier molecules (monolayer). The particle size is variable and usually covers the range 5–10 nm (micelle), 20–50 nm (microemulsion or swollen micelle), 100 nm (nanoemulsion), 50–200 nm (solid lipid nanoparticle, SLN) to a very large 500–5000 nm (coarse emulsion) (Becher, 2001). At sizes above 5 microns, dispersed droplets are relatively unstable and are not suitable for injection (as the capillary diameter of vasculature may be as small as 20–30 μm), but they may be used externally for topical preparation. Still larger droplets lack usefulness in pharmacy as they are susceptible to creaming, followed by increased rates of coalescence (Sarker *et al.*, 1998a,b).

Figure 2.1 shows the three basic forms of coarse emulsion. Emulsions that are primarily composed of spherical droplets can be defined by a phase volume (φ). Particles remain spherical up to a maximum φ of 0.74; in this case, spheroids assume a polyhedral appearance and are present as a highly concentrated emulsion. Upon loss of spherical shape, many of the droplets undergo coalescence or phase inversion. In pharmacy and cosmetology, emulsions are frequently used to place the apolar drug dose in a given specific volume. These are usually oil-in-water (O/W) emulsions. The oil and water form the dispersion phase (dependant on the emulsifier hydrophile–lipophile balance (HLB) or a mix of emulsifiers and overall HLB). The form and size of the emulsifier alkyl chain determine both its HLB and thus the type of emulsion which forms. The phase that forms the dispersed or continuous phase also depends on the pharmaceutical formulation and temperature.

Given appropriate mechanical energy input, dispersed droplets of oil (or water) form emulsions. These emulsions may be called creams, ointments, liniments (balms), pastes, bases or occlusive films or liquids, depending mostly on their oil and water proportions and their use. Topical dosage forms (e.g. creams and

Pharmaceutical Emulsions: A Drug Developer's Toolbag, First Edition. Dipak K. Sarker.
© 2013 John Wiley & Sons, Ltd. Published 2013 by John Wiley & Sons, Ltd.

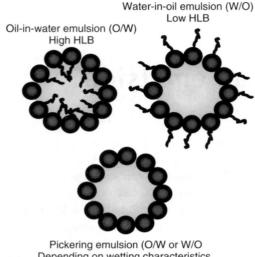

Figure 2.1 The anatomy of the 'course emulsion', often referred to simply as the 'emulsion'. Course emulsions typically have diameters larger than 500 nm, usually 1–2 µm. HLB, hydrophile–lipophile balance

ointments) are used on the surface of the skin, transdermally, vaginally or rectally. Liquid-like ($\varphi = 0.1 - 0.2$) emulsions can also be used orally or injected via various routes (typically intravenously or intramuscularly). Frequently encountered topical emulsions include acyclovir (cold sore) cream, hydrocortisone cream and clotrimazole vaginal antifungal cream.

A micelle (see Figure 1.2) is a self-assembled emulsifier-based particle (see Sections 2.1.1, 4.3, 5.2.5, 7.2, 7.4, 8.3, 12.2 and 12.4), used to deliver high log P apolar drugs such as miconazole (Sarker et al., 1995a,b; Sarker, 2005a,b, 2006a), which is used in the treatment of thrush. Micelles form spontaneously and are thermodynamically stable; like microemulsions, they can only form in the presence of polymeric or simple emulsifiers. A 'swollen micelle' or microemulsion (Figure 2.1) is used to deliver apolar drugs in 'solution' for ocular therapy (e.g. anaesthesia) and to encapsulate vaccines. Typically, the emulsions are nanoemulsions of soybean (soya) oil, with particle diameters of 20–50 nm. They form spontaneously but require two or more emulsifiers of varying sizes and much higher concentrations of surfactant (10–25% w/v emulsifier) than for conventional coarse emulsions. They can be manufactured using synthetic or naturally occurring emulsifiers such as bile salts (sodium deoxycholate, taurocholate). Microemulsions, having a lipid-filled core, can thus be used to fabricate SLNs.

Nanoemulsions (Sarker et al., 1999; Sarker, 2005a,b, 2006a; see Figure 1.2) are smaller versions of coarse emulsions and are phenomenologically unrelated to microemulsions. They are typically 50–400 nm in diameter and are limited to a minimum size of about 40 nm, due to surface coverage. They are produced by extremely high-shear mixing of dilute emulsion premixes using strongly surface-stabilising mixtures of lipids, low-molecular-weight (LMW) emulsifiers and polymeric emulsifiers (Sarker, 2005b, 2006a). Some types of nanoemulsion are currently under study for their use against HIV-1 and TB pathogens (Richards *et al.*, 2004).

Liposomes (see Figure 1.2) represent another form of lipid encapsulation (Sarker, 2009a). They consist of a bilayered envelope (lamella, bilayer, leaflet) with an aqueous core. Since a micelle or liposome is a lipid (apolar) based structure and is made of 'fat', solubilisation of the drug in an alkyl chain-rich environment is analogous to a form of emulsification. In this book, the term 'emulsification' is used in the 'true sense' but also to infer encapsulation and lipid solvation. Some unusual forms of emulsion can be fabricated using liposomes. One such example is the use of liposomes (a form of dispersion) to fabricate multiple emulsion droplets (Wang *et al.*, 2010). In this case, their use as an immunological adjuvant induces the production of cytotoxic T-lymphocytes raised against an HIV envelope antigen (Richards *et al.*, 2004).

Such fascinating complexity is more often exploited in a simpler form, for example as:

- Vesicles or liposomes.

Many modified liposomes are currently under investigation (Sarker, 2009a). Basic types include:

- Small unilamellar types (SUV), large unilamellar types (LUV) and multi-lamellar types (MLV). Depending on the type of component used in the fabrication, these are routinely known as:
  - proteoliposomes: constituted with antigens;
  - chitosomes: constituted with chitosan;
  - niosomes: constituted with nonionic surfactants;
  - dermosomes: constituted with phospholipids and cholesterol;
  - virosomes: constituted with viral antigens;
  - polymersomes: constituted with polymers.

In any case, the three basic components of all dispersions (including liposomes) based on oil are water, oil and emulsifier(s), but other components may be included, such as drug, antioxidant and preservatives.

Droplets which make up emulsions (creams, lotions, ointments, nano- or microemulsions) are sometimes referred to as particles (see Figure 1.2 and Figure 2.1). These particles are separated by a thin liquid film (TLF) or lamella. The lamella terminates at an intersection called the Plateau border (or a node where Plateau borders meet). The TLF, its surface chemistry and its mechanical properties play an often understated role in determining suspension and emulsion metastability (see Sections 2.2.1, 3.2, 3.3 and 14.3).

One prerequisite for the formation of any emulsion is that it makes use of an emulsifying agent to form particles; without this, coalescence and liquid droplet fusion are inevitable. Even with SLNs (solid emulsions), initial formation is impossible without the presence of an emulsifier (Sarker, 2012a). The dispersed phase in continuous phases (dispersion medium) determines the nomenclature. There are three basic forms of coarse emulsion: O/W, water-in-oil (W/O) and multiple (e.g. water-in-oil-in-water). The most common forms in pharmacy are O/W emulsions, particularly for parenteral administration; W/O forms do exist, but they are normally reserved for dermal applications. Use of emulsions can be for taste-masking purposes but is usually for solubilisation of apolar moieties. For all emulsions, the increase in energy associated with dispersal makes the system thermodynamically unstable; however, the discrepancy is filled by a monolayer or multilayer of adsorbed material (i.e. surfactant/emulsifier, hydrocolloids, finely divided solids (see Section 2.3.2)).

True colloids (microemulsions, SLNs, micelles) are defined by a nanometre size, by being thermodynamically stable and by spontaneous formation. Coarse emulsions are not colloids, although they may have components which are colloidal (e.g. adsorbed macromolecules), but rather are dispersions (mostly opaque), generally being of a micrometre size (or bigger) range. Emulsions typically have higher surface free energy than microemulsions (swollen micelles) and cannot be reconstituted.

The use of emulsions for creams, ointments and parenteral (injectable) medicines is discussed at length in Chapters 5 and 7. One dramatically expanding area of emulsion use is as vaccines (Richards *et al.*, 2004), or rather as adjuvants to promote/enhance immunological effect. Examples of such uses are presented in Table 2.1.

Emulsion vaccines frequently use biocompatible fatty acids and lecithin-based polymer derivatives as emulsifiers, such as Transcutol P (Gattefossé), Kolliphor EL (BASF) or Miranol C_2M (Rhodia). Therapeutics of this type are not new, since the first therapeutic based on oil, which used cyclosporine and Cremophor, was approved by the US Food and Drug Administration (FDA) in the 1980s. Some interesting examples of in-development (Table 2.1) and developed vaccine emulsions are presented in Figure 2.2.

The recipes for internal versus external application of emulsions are similar and generally use the same generally-recognized-as-safe (GRAS)- and

Table 2.1 Emulsions used as vaccines and vaccine aids

Emulsion type	Vaccine target	Type of dispersed phase	Example of oil/fat
O/W	Haemorrhagic fever,[a] influenza, Ebola, rabies, smallpox, tumour antigens, autoimmune, melanoma, HIV	Mineral oil, e.g. paraffin	Montanide ISA 51
O/W	HIV, malaria	Nonmineral oil	Montanide ISA 720
	Pneumococcal, rotavirus, five-in-one vaccine (diphtheria, tetanus, pertussis, polio, *Haemophilus influenzae*)	Nonmineral oil	Merck '65-4' + arlasel aluminium; arachis (peanut) oil base
	HIV, influenza, *Herpes simplex*, hepatitis B, malaria, cancer	Nonmineral oil	OM-Pharma OM-174 (mono-phosphoryl lipid A); Antigenics QS-21 (acylated saponin)
W/O	Topical infection, veterinary applications, e.g. foot-and-mouth virus	Mineral oil	Montanide ISA 25D
W/O/W	Snake venom vaccine,[a] veterinary applications	Mineral oil	Montanide ISA 206D
Solid lipid nanoparticle (O/W)	Anthrax, cholera, typhoid	Nonmineral oil	Tristearin

Nonmineral oils include Vegetal (peanut, sesame, olive, soya), animal oils and squalene (Novartis' proprietary O/W emulsion adjuvant MF59 (O/W squalene nanoemulsion platform—AddaVax)). O/W, oil-in-water; W/O, water-in-oil; W/O/W, water-in-oil-in-water.
[a]Waghmaree *et al.* (2009).

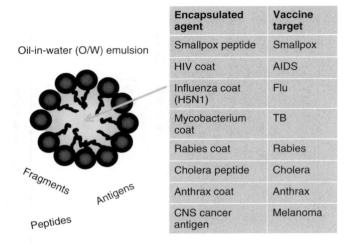

Encapsulated agent	Vaccine target
Smallpox peptide	Smallpox
HIV coat	AIDS
Influenza coat (H5N1)	Flu
Mycobacterium coat	TB
Rabies coat	Rabies
Cholera peptide	Cholera
Anthrax coat	Anthrax
CNS cancer antigen	Melanoma

Oil-in-water (O/W) emulsion

Fragments

Antigens

Peptides

Figure 2.2   Use of emulsions as adjuvants and as vaccines for therapeutic isolation of immune-responsive antigens and fragments

biocompatible-grade emulsifiers, fats and oils. The oil usually accounts for 10–90% of the sample and depends on internal and external usage. Vaccine emulsions are usually formulated as a W/O or an O/W format, depending on the primary excipients used. Most emulsions for injection are formulated to have a low phase volume of the dispersed phase of around 0.1, due to the lower viscosity (see Table 2.2). The 'standard' formulation for an emulsion would be:

- water (often referred to in trade products as 'aqua');
- liquid paraffin/paraffin wax/fish oil/vegetable oil (oil phase);
- beeswax/microcrystalline wax (structuriser);
- sodium benzoate (E216)/preservative;
- propylene glycol (thickener, water binder);
- citric acid/sodium phosphate (buffer);
- vitamin E/ascorbyl palmitate/butylated hydroxyanisole (BHA) (antioxidant);
- sorbitan sesquioleate (emulsifier/structuriser/synergistic lipid);
- tragacanth gum/modified cellulose gum (thickener/stabiliser) for external use, along with high-phase-volume O/W format.

Coarse emulsions always require appropriate high-energy (shear) mixing and the inclusion of a drug/active appropriate for the application. An example would be the 1% w/v propofol emulsion for intravenous (IV) injection for anaesthesia.

Table 2.2 Types of crude dispersion and colloidal (submicron) dispersed system

Part I: key delivery systems

Topical	Creams, ointments, lotions, emollients, absorption bases, soaks and pastes
Transdermal 'devices'	Patches, implants, dressings, pessaries, suppositories
Parenteral	Nutrition, imaging, radiotherapy, chemotherapy, anaesthesia, vaccines
Oral	Specialised GIT, vitaminised products, soft gelatin capsules

Part II: types of 'emulsion' dispersion

Type	Dispersed phase (DP)	Continuous phase (CP, continuous medium)	Surface stabilisation	Size (diameter)
Basic: O/W	Oil	Water	Higher HLB emulsifier	1–5 μm
Basic: W/O	Water	Oil	Lower HLB emulsifier	1–5 μm
Multiple: O/W/O	O/W	Oil	Higher and lower HLB emulsifier, respectively	15–500 μm
Multiple: W/O/W	W/O	Water	Lower and higher HLB emulsifier, respectively	15–500 μm
Microemulsion (swollen micelle):[a] O/W	Oil	Water	Two emulsifiers needed to reduce interfacial tension; primary is higher HLB	20–50 nm
Microemulsion (swollen micelle):[a] W/O	Water	Oil	Two emulsifiers needed to reduce interfacial tension; primary is lower HLB	20–50 nm
Nanoemulsion: O/W	Oil	Water	Higher HLB emulsifier	50–200 nm
Nanoemulsion: W/O	Water	Oil	Lower HLB emulsifier	50–200 nm

[a]Form spontaneously (requires use of multiple emulsifiers, e.g. primary and coemulsifier, often an alcohol). Nomenclature: higher means ∼9–12 and lower means ∼3–6. O/W, oil-in-water; W/O, water-in-oil; O/W/O, oil-in-water-in-oil; W/O/W, water-in-oil-in-water; HLB, hydrophile–lipophile balance. Surfactant micelle is 2–5 nm diameter; polymeric micelle about 50 nm. Microemulsion and nanoemulsion are not the same. Microemulsions are thermodynamically stable, all other emulsions are metastable.

The manufacturers emulsify the fat-soluble propofol (anaesthetic) in a mixture of water, soya oil and egg lecithin.

The basic colour of all coarse emulsions made from moderately sized (several–many micron) droplets is an opalescent or opaque white (Sarker, 2005b), due to scattering of light and the Tyndall effect. This is modified to blue-white if dilute and yellow-pink if concentrated. Coarse emulsions made from small, 500–1000 nm (0.5–1.0 μm), droplets appear greyish, whereas microemulsions and 'related nanoemulsions' tend to appear colourless (optically transparent) due to the small size of the dispersed (disperse) phase and limited interference with the simple passage of light through the sample.

For coarse (nanosized) emulsions, energy input is through vigorous shaking, stirring, high-shear (MPa stresses and kHz rates) homogenising or spray processes, which are necessary for initial formation of an emulsion. Emulsions show a tendency to cream and coalesce and consequently 'split', reverting to the stable state of the 'two' distinct phases composing the emulsion on standing. An example of this is the separation of liquid paraffin and water typically seen in adsorption bases (e.g. paediatric emollients), which will quickly crack or split (separate) unless shaken repeatedly.

As pharmaceuticists, we talk frequently about the stability of W/O or O/W emulsions, but a coarse dispersion is inherently metastable. In reality, an emulsion's 'stability' simply refers to its ability to resist change in its 'appearance' over an extended period (days, months or years—usually). All emulsions are really 'unstable' with respect to time. It is however possible to quantify a product's 'relative stability', hereafter referred to as 'stability', as represented in Figure 2.3. In this approach, it is entirely feasible (as shown) to quantify the product's relative stability using the aqueous conductance of buffering ions (Sarker et al., 1999).

According to the Institute Nationale de la Recherche Agronimique (INRA) creaming apparatus, as creaming (floatation caused by buoyancy effects) of droplets occurs on a large scale, the subnatant and supernatant change physical compositions (in terms of phase volume), revealing an increase in conductivity of the increasingly water-rich subphase.

The most common way of investigating shelf life is by the use of certain thermal methods, which consist in increasing the temperature in order to accelerate destabilisation (Dimitrova et al., 2000) by augmenting particle collisions and exaggerating density gradients within the sample. This also has the effect of locally decreasing interfacial viscoelasticity (below critical temperatures of phase inversion and chemical degradation). Temperature affects not only the viscosity (via the PIT), polymer solubility (Krafft temperature) and internal friction, but also interfacial tension, and can thus drive coalescence or phase inversion (Sarker, 2005b). Mechanical acceleration, including vibration, centrifugation and agitation (perikinesis), is sometimes also used to test shelf life by driving flocculation (see Section 3.2.1) and creaming (Manoj et al., 1998; Dimitrova et al.,

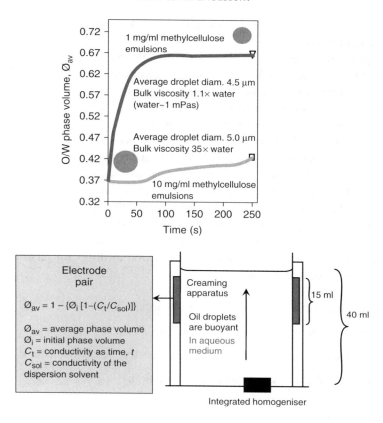

Figure 2.3 Schematic representation of conductimetric methods of examining stability factors such as the creaming and coalescence of emulsions. One such system was developed at the Institute Nationale de la Recherche Agronomique (INRA), Nantes, France. Such systems permit any sample creaming to be followed by measurement of the conductivity of the aqueous subnatant as the supernatant becomes more laden with the less dense oily phase

2000). Differential segregation of different populations of particles is seen during centrifugation and certain forms of intense vibration.

An emulsifier (also sometimes known as an emulgent) provides protection against flocculation in terms of steric constraint and energetic (see Sections 2.2, 2.3.2, 3.2 and 3.3) restrictions, and stabilises emulsions by increasing their avoidance of coalescence. A range of emulsifiers are used in pharmacy, from small aliphatic molecules (e.g. fatty acids) to lipids (e.g. soybean lecithin) and derivatised polysaccharides and mucilages such as alginate, hydroxypropyl- and hydroxyethyl-celluloses (HPC or HEC) and tragacanth. Solids such as carbon (nanocarbon), silicates, organic nanoparticles and latex particles can also stabilise emulsions, through a mechanism called Pickering stabilisation (Pickering, 1907; Binks, 2002). Parenteral emulsions are biocompatible O/W emulsions which are customarily stabilised with egg yolk/soya lecithin with vegetable oil (arachis, castor) or liquid paraffin.

Amphiphilic polymers are another class of emulsifier. They physically interact with both oil and water, stabilising the interface between oil or water and droplets in a coarse suspension. A plethora of emulsifiers are used in pharmaceutics to fabricate emulsions, such as creams and ointments. Common topical and medicinal examples include emulsifying wax, wool fat–lanolin (cosmetics), oleyl alcohol derivatives, cetostearyl alcohol derivatives, polysorbate 20, cetomacrogols and cetosteareth 20. Sometimes the particles (and fat crystals) from the inner phase can migrate to the interface and themselves act as an emulsifier, and the result is a 'solid' stabilised emulsion (Binks, 2002; Concannon *et al.*, 2010). This has been proposed as a new form of controlled drug delivery. It is also possible to use cosolubilisation aids such as poly(ethylene glycol) (PEG), which can facilitate sequestration and solubilisation and therefore aid constitution.

In order to ensure a fuller emulsion solubilisation of drugs for pharmacy products, formulators should consider:

- Maintaining 'sink' conditions that drive dissolution.

- Oil solubility/lipophilicity, where drug log P <0.8.

- The location of the drug within the product (this is known as is its 'form').

- Avoidance of drug hydrolysis and flocculation.

2.1 States of matter

There are three basic states of matter: solids (gel phase in the case of a fat), liquids and gases. However, many variations are also possible (Figure 2.4a), such as the various liquid crystalline (LC) states that lipids and fats exist in, including nematic, smectic, cholesteric phases (Figure 2.4b). Lipids and surfactants also exist in LC states, such as lamellar, cubic, hexagonal and continuous (or wormlike). Consequently, both the oil phase (or emulsion/microemulsion) and the emulsifier can exist in a range of physical states (Jones, 2002). This has implications for the 'stability' and physical form of the product.

The relationship between state and function become even more significant in a range of novel materials for pharmaceutical use (Sarker, 2010, 2012a). These biomaterials can include simplified composite-controlled release matrices, inorganic and organic alloys, solid foams (for subcutaneous implantation) and the customary range of semisolids (gels and thickened or concentrated emulsions or dilute emulsions, where φ <0.4 (viscosity <100 mPas)). Gases, such as pentaflourooctadecyl (Oxygent) bromide (used as a blood substitute, in the form of an aphron), and a range of other fluorinated materials can also be used to

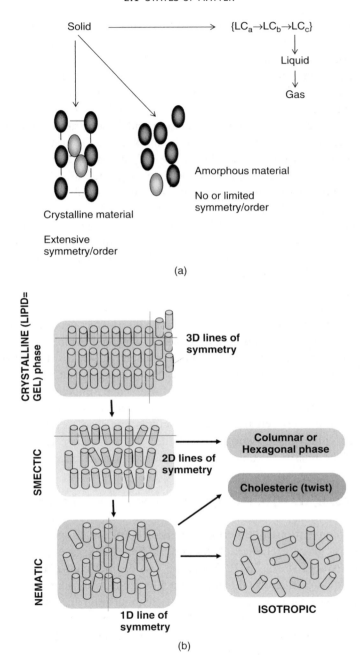

Figure 2.4 Simplification of the principal states of matter for fats and oils, polar lipids, surfactants and polymers used to create emulsions, liposomes and micelle-based therapeutics. The forms are solids, liquid crystals (many subtypes), isotropic liquids and gases. Liquid crystals can have one-, two- or three-dimensional symmetry, producing a variety of textures. (a) Overview. (b) Some of the types of liquid crystal

create nanobubbles/microemulsions (aphrons) and coarse emulsions for medical and biomedical imaging applications. These materials show a change of state, which can be used favourably (see Section 5.2.5).

Conventional pharmacy LCs (mesophases) can be divided into thermotropic and lyotropic phases. Thermotropic LCs exhibit a phase transition into the various LC phases as temperature is changed. Lyotropic LCs exhibit pronounced distinct phase transitions as a function of both temperature and concentration change within a solvent (typically water). Phase or state is important, as it influences the one- to three-dimensional symmetry (1–3D) of the LC observed, and therefore its use. The forms most commonly seen are proper (isotropic) liquids, 'threadlike' nematic phases (1D symmetry), 'greasy' smectic types of many subtypes (2D symmetry) and intact solid crystal forms (3D symmetry).

Lyotropic LCs consist of two or more components (which require an amphiphilic molecule; HLB drives the prevailing form), which exhibit LC properties in certain concentration ranges. In the lyotropic phases, solvent molecules fill the space around the compounds, providing fluidity to the system. In contrast to thermotropic LCs, these lyotropic forms have another degree of freedom of concentration, which enables them to induce a variety of different phases. Lyotropic forms can be isotropic fluids, micelles (low concentrations), vesicles (bilayered micelles) and hexagonal-columnar-lamellae, which are observed at higher concentrations. These anisotropic self-assembled nanostructures can order themselves, much like solid-based LCs, leading to large-scale versions of all the thermotropic phases (e.g. a nematic phase of rod-shaped micelles). Other systems, such as cubic phases, may exist between the hexagonal and lamellar phases, and interconnectivities between individual structures can lead to the formation of a bicontinuous cubic phase (wormlike micelle). At yet higher concentrations, an inversion of the phases (hexagonal, cubic/micellar) can occur.

2.1.1 Types: O/W, W/O, multiple, microemulsions, nanotypes

The array of conventionally described emulsion forms, sizes, polydispersities and complexities is given in Table 2.2 (Sarker, 2005b; Araujo *et al.*, 2007; Cevc and Vierl, 2010; Araujo *et al.*, 2011). There are two distinct classes of emulsion, but many forms of presentation (e.g. liquid for injection or transdermal patch). Principally, there are coarse or crude emulsions (without inferring inferiority). These can be subdivided into large (micron size and nanoemulsions) and fine or thermodynamically stable forms, which include microemulsions and various forms of micelle. Newer forms can now include spray-dried emulsions and solid-in-oil (in-water/in-solid) emulsions. Uniformity of product (gels, solutions, suspensions (solid-in-water, S/W), semisolids (e.g. O/W emulsions) and pastes (powders), all based on immiscible fat and water mixtures) determines both the efficacy (performance) and the shelf life of the dosage form.

Certain parameters define the correct function and consistency of any emulsion product. These are:

- Creaming and rates of creaming (favouring coalescence on compaction).

- Particle size and charge permitting flocculation (and thus coalescence).

- Hysteresis and weeping in concentrated >0.74 phase volume emulsions.

- Consistency/viscosity and its impact on the rate of particle movement and thus collision.

- Thixotropy (shear-induced thinning) and the rheodestruction that might occur on the application of the product (e.g. topical application to the skin).

- Flocculation of particles due to apolar emulsifier-facilitated or charge-driven interparticle interaction.

For products stored on the pharmacy shelf, it is possible to model flow behaviour and its impact on quality with standing. The following expression applies:

$$\eta = \exp{}^{(kB/T)} \tag{2.1}$$

Any suitable stability assessment would need to consider Ostwald ripening (droplet growth driven by polydispersity), phase inversion and other changes in the dispersed phase based on temperature, and the impact these could have on particle size. The storage time and temperature are crucial in determining stability. In addition to bulk properties, they shows themselves as biphasic separation of the emulsion surface adsorbed layer of emulsifier and its phase (mechanical) change, which can also modify product stability. Manufacturers of emulsion products examine and consider the following both for evaluation of the risk of product change and to formulate an idea of product longevity:

- temperature;

- humidity;

- lighting intensity cycling.

As well as their effects on degeneration.

The product is assessed in terms of accelerated testing regimes, using the following indices of quality:

- particle size/ ζ-potential;

- rheology (viscosity, viscoelasticity, 'hardness', etc.);

- visual appearance;

- drug content;

- microbiological load (e.g. bacterial lipolysis and lipase action);

- product uniformity;

- melting point and crystal form (differential scanning calorimetry, DSC; see Sections 2.1.2, 6.2, 6.3 and 12.1).

The product is made more unstable by:

- physical change, e.g. polymorphism;

- chemical change, e.g. pH increase or decrease;

- microbiological change, e.g. biosurfactant and mucilage generation.

One factor that clearly does drive coalescence is compaction and compression of droplets in the creamed layer of a standing emulsion. This compaction is exaggerated and augmented by interparticle attraction, which is influenced by:

- The extensivity of the droplet palisade layer of adsorbed emulsifier.

- Polydispersity and particle packing.

- Mechanics (surface viscosity and how this is changed by the nature of the adsorbed material).

- Desorbed polymer and weak, intermittent or sporadic surface coverage on the particle surface.

- Interaction between particle and continuous phase via hydrophilic emulsifier coating (e.g. poloxamers and methylcelluloses).

The nature of the palisade layer of an emulsion, microemulsion, multiple emulsion, nanoemulsion or polymeric micelle is shown in Figure 2.5. The surface-constrained polymer and its interaction with neighbouring molecules and solvent play a key role in inter- and intraparticle association and stabilisation.

The palisade layer can be described as the surface free-standing chains protruding away from the body of an oil droplet (particle) as a result of emulsifier adsorption on to the surface. Its orientation depends on surface pressure (loading; $\pi = \gamma_{water} - \gamma_{surfactant}$) and can exist in a monolayer or multilayer form, altering the particle surface viscosity.

Microemulsions (W/O or O/W forms) can also possess a palisade layer and, although they are seldom described in exemplary detail, they do show great promise as therapeutic vehicles, despite requiring large emulsifier concentrations. Microemulsion samples are usually optically transparent, monodispersed, with low viscosity and a particle diameter that is normally 10–50 nm (but can reach up to

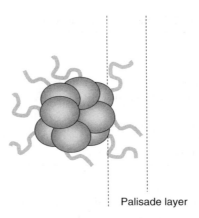

Palisade layer

Figure 2.5 Representation of the general form of a polymeric micelle. The inner portion or core comprises the apolar (lipophilic, lipidic) portions of the emulsifying polymer; these give the core some characteristics that resemble an oil phase. The outer portion is referred to as the palisade layer (of protruding hydrophilic/polar chains). The extensivity of the polar region can confer some degree of chemical protection and long-circulating characteristics for injectable drug forms

200 nm), as described in Table 2.2. They also have a phase volume (φ) of 0.2–0.8, but require an emulsifying adjunct (cosurfactant) so that interfacial tension (γ) can be reduced to 0 mN/m or less (permitting spontaneous emulsification). In a similar manner to a nanoemulsion (a small version of a coarse emulsion), they are small enough to avoid clearance in the liver (monocyte–phagocyte system, MPS; see Chapter 7 and Sections 12.3 and 12.4) and so are ideally suited to parenteral/depot or ocular use. Microemulsion stability as a result of constitutional change in these environments underlies their limited general use (Sarker, 2005b, 2006a; Lawrence and Rees, 2012). Additionally, their primary disadvantage is that they need much more emulsifier ($\sim$25%) than a micelle. Today, microemulsions are used routinely for or are being investigated for use in:

- Encapsulation of flavours and pigments for pharmaceutical syrups.

- Drug delivery, such as chemotherapy (e.g. cytotoxic methotrexate diester in W/O forms) and analgesia (indomethacin).

Emulsions, nanoemulsions and microemulsions represent the traditional form of the emulsion particle. Liposomes, micelles, LCs based on dispersed colloids and SLNs also represent a form of 'emulsion'. Table 2.3 shows the 'molecular ruler' of sizes traversed in these types of lipid-based dispersed pharmaceutical product (Primo *et al.*, 2007; Davis *et al.*, 2008).

The size of emulsion is important as it governs both colloidal stability and drug and/or particle uptake and efficacy *in vivo* (Valtcheva-Sarker *et al.*, 2007;

Table 2.3   Basic dimensions of nanoparticles

Particle	Diameter (nm)
Microsphere	200–1000
Multilamellar vesicle (MLV)	200–1000
Solid lipid nanoparticle (emulsion)	50–200
Nanocapsule	50–200
Small unilamellar type/nanoparticle	25–200
Microemulsion	20–50
Surfactant micelle	5–10

Kaur *et al.*, 2008; Sarker, 2012a). Additionally, for particles entering the blood, circulatory persistence is augmented when the particle size is kept below 50 nm (Sarker, 2006a; Valtcheva-Sarker *et al.*, 2007). Targeting the brain and nervous system with nanoparticle vehicles is made more difficult by the inherent impermeability of the structure, the transmembrane transportation and the tight junctions of this tissue. This impermeability is often referred to as the blood–brain barrier (BBB) and poses one of the greatest challenges to the drug developer (Hu and Dahl, 1999; Seelig, 2007; Maeda *et al.*, 2009). Tight junction sizes ranging from 3 to 15 nm have been reported; molecules and particles of this size (occasionally subject to *in situ* modification) are limited to molecular masses of around 250 kDa. This greatly favours the use of microemulsions and micelles rather than 100 nm vehicles for targeting. This size barrier need not be inhibitory, however, as it can be circumvented in some cases, since an anisotropy is also manifested in the interfacial energy (surface tension) between different LC phases, as seen with spread monolayers of micelles and microemulsions, which means they are more likely to disintegrate in a particular environment (Lawrence and Rees, 2012). Complex mixtures of varying types and concentrations of cosurfactants and oils allow for the ability of emulsifiers to efficiently cover droplet surfaces (e.g. PEG-8 caprylic/capric glycerides (Labrasol), cosurfactants (polyglyceryl-6 dioleate (Plurol Oleique)) and polyglyceryl-6 isostearate (Plurol Isostearique)) and can be used to engineer surface phase behaviour and fusogenicity or to modify *in situ* performance (Sarker, 2012a).

Nanoemulsions are nanosized coarse emulsions. Some scientists continue to confuse them with microemulsions: microemulsions are indeed nanosized but are energetically distinct by virtue of being more appropriately assigned as 'swollen micelles'. The interest in nanoemulsions started in the 1960s with Kabi (Pharmacia) in Sweden, who developed Intralipid for parenteral nutrition (70–400 nm). Their recipe used soybean oil, egg yolk phospholipid (phosphatidylcholine (PC) and some other phospholipids), emulsifiers and glycerol. Other than nutrition, the delivery form has since been used (Sarker, 2008) for propofol (Diprivan), diazepam (Diazemuls), a flurbiprofen prodrug (Lipo-NSAID) and emulsions used in imaging (e.g. lipiodol CT and Fluosol/Oxygent (see Chapters 7 and 11 and Section 5.2.5)). The efficacy of nanoemulsions can be limited, as permeation in many cases is restricted by drug lipophilicity (log P) for transcellular passage or by tight junction geometry (<500 kDa) for paracellular drug delivery (Hu and Dahl, 1999; Seelig, 2007; Maeda et al., 2009).

2.1.2 Fats and oils

The fats and oil used in pharmacy products have to be biocompatible for use *ex vivo* and *in vivo* for a multitude of heavily processed products (Huailiang et al., 2001). All oils (and solid fats) must usually remain nontoxic and be readily excreted or metabolised as a result of administration (Florence and Attwood, 1998). For this reason, there isn't an exhaustive list of potential candidate (GRAS) oils, but rather an extensive list of approved and regulated natural and synthetic (e.g. paraffin, silicone) oils (Sarker, 2005a). The following oils and fats are the most widely used:

- *Common naturally occurring oils* Castor, soya, arachis, safflower, squalene (cosmeceuticals), olive, sunflower, evening primrose oils, fish oils, esterified oils.

- *Common fats (for cosmetology)* Tallow, shea butter, lanolin.

- *Common fats (for pharmacy)* Tristearin, tri-/di-/monoglycerides, cocoa butter, hydrogenated plant/vegetable oils (e.g. coconut oils), esterfied fats.

- *Synthetic 'oils'* Vaseline/petrolatum or paraffin (light or heavy), simeticone/dimethicone.

Obtaining oil of the appropriate chemical make-up and purity, both cheaply and reproducibly, is an important consideration for all large-scale drug product-development teams (Sarker, 2008).

Table 2.4 highlights the fact that excipients and vehicles are subject to compositional change. This is manifested clearly in the defined melting points for cocoa butter (Table 2.4) and oleic acid. Some oils, such as castor oil (hydroxyl-group 'rich'), make the product stable. For enteric emulsion and parenterals, 'stable', nontoxic, nondigested oils are required. Other oils, such as paraffin, are alkane oils with light to heavy fractions (low and high viscosity, respectively) and a general formula C_nH_{2n+2}. Paraffin 'wax' refers to a mixture of alkanes with $20 \leq n \leq 40$ sizes. They are solid at room temperature and begin to enter the liquid phase at approximately 37°C. Paraffin waxes' melting points lie between 40 and 65°C and

Table 2.4   Commonly used fats and oils with key physicochemical properties

Type of solid fat or oil	Melting point, T_m (°C)	Density (g/cm³)	Mainly saturated chain	Primary fatty acid (%)
Tristearin	55–74	0.86	✓	Stearic (100)
Cocoa butter[a] (theobroma oil)	34.0–36.5	0.96	✓	Stearic (35) Oleic (35)
Squalene	−75	0.86	No	Triterpene only (100)
Butter fat	<40	0.91	✓	Palmitic (30)
Lanolin	42	0.95	✓	Lanolic C_{12}–C_{24} (97)
Olive	−6	0.92	No (<10% saturation)	Oleic (75)
Arachis (peanut)	2	0.91	No	Oleic (50)
Soya	−16	0.93	No	Linoleic (65)
Castor[b]	−18	0.96	No	Ricinoleic (85)
Oleic acid	13–14	0.9	No	Oleic (100)
Paraffin	37	0.9	✓	C_{20}–C_{40} alkane

[a]10% dispersion has a pH ~5.7.
[b]Has a viscosity of about 500 mPas at room temperature (water is 1.0 mPas).

they have a density of around 0.9 g/cm³. They are insoluble in water and are unaffected by most common chemical reagents. They are thus ideal for external or suppository use. Other favoured oils include:

- arachis (peanut);
- soya (soybean);
- palm;
- olive;
- safflower;
- sunflower;
- poppy seed;
- corn (maize).

For some products, such as interfacial destabilisers and defoaming agents (also breast implants, cosmetics), silicone oils (e.g. dimethicone, simeticone, poly(dimethylsiloxane) (PDMS)) are widely used. Silicone oils (density of around 0.8 g/cm³) have the general formula $CH_3[SiO(CH_3)_2]_n Si(CH_3)_3$, where n is the number of repeating monomer $[SiO(CH_3)_2]$ units. Mineral oil has a number of names, including adepsine oil, Glymol, Alboline, Nujol, medicinal paraffin and Saxol. Additionally, certain iodised vegetable oils and animal lipids are used for products designed for BBB drug delivery.

Table 2.5 Polymorphic forms of theobroma cacao (cocoa butter) and associated melting point classifications, discovered over several decades

Vaeck model, 1960		Duck model, 1964		Willie and Lovegren model, 1966		Lovegren et al. model, 1976		Davis and Dimick model, 1986	
Form	Temp ($^{\circ}$C)	Form	Temp ($^{\circ}$C)	Form	Temp ($^{\circ}$C)	Form	Temp ($^{\circ}$C)	Form	Temp ($^{\circ}$C)
γ	17	γ	18	I	17.3	VI	13	I	13.1–17.6
α	21–24	α	23.5	II	23.3	V	20	II	17.7–19.9
				III	25.5	IV	23	III	22.4–24.5
$\beta\prime$	28	$\beta\prime\prime$	28	IV	27.3	III	25	IV	26.4–27.9
		$\beta\prime$	33	V	33.8	II	30	V	30.7–34.4
β	34–35	β	34.4	VI	36.3	I	35.5	VI	33.8–34.1

Suppositories are oleaginous (triglyceride) bases, such as theobroma oil/hydrogenated hard fat coconut palm oil or base (e.g. Witepsol/Suppocire; Chaiseri and Dimick, 1986) and were used for significant numbers of drug delivery systems (DDSs) in the past. Table 2.5 shows a recent classification of theobroma oil (cocoa butter). Although apparently simple, it emphasises the difficulty for the product formulator in terms of excipient polymorphism and the net effect on product PCQ in the event of physical change.

When choosing a base, the one in which the drug is least soluble is selected; the product is often more of an amalgam than a traditional emulsion (e.g. many oily phases used in transdermal patch technologies). Formulation of a base (Chaiseri and Dimick, 1986) supposes that:

- It melts at or close to body temperature (and/or dissolves in body fluids, e.g. rectal suppositories or vaginal pessaries (suppositories for vaginal delivery)).

- It is both nontoxic and nonirritant in nature.

- It is compatible with any therapeutic.

- It releases any drug freely and without hindrance.

One strategy for dealing with polymorphism in the solids fats used in SLNs, suppositories and pessaries involves warming the fat. Heating to $50 - 52^{\circ}$C eliminates the physical 'memory' of fat polymorph crystals; subsequent cooling to 25°C then establishes the type I–III forms (Table 2.5), while tempering or holding of the room-temperature-cooled fat at $33 - 35^{\circ}$C allows form IV to predominate. Using Span 60/65, Tween 60/65 or propylene glycol emulsifiers at low concentrations in the sample maintains forms IV and V as a liquid fraction and eradicates the unsavoury 'gritty bloom'.

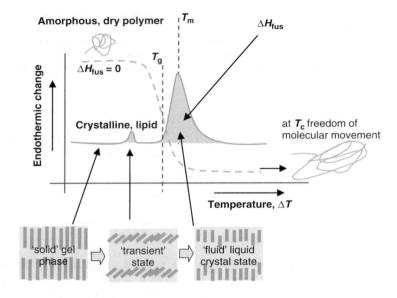

Figure 2.6 Thermal properties and transitions in lipids and fats, and the polymers used to create encapsulated drug delivery particles. When studied using differential scanning calorimetry (DSC), lipids have a melting point (T_m) (and a pretransition melting) as they move from gel phase to liquid phase, whereas polymers show a glass transition temperature (T_g), typically associated with chain movement

Submicron drug delivery lipid nanoparticles are even more complex as they combine fat and polymeric emulsifier or a PEG 'stealth' coating. The typical thermal behaviour of these products is represented in Figure 2.6. Many fats show a premelting transition, clearly visible in the figure, and this can be falsely assumed to be due to an impurity or the actual endothermic melting peak. Polymeric materials form amorphous phases and show a glass transition temperature (T_g) rather than a melting temperature (T_m); additionally, enthalpy for the glass transition is zero (Jones, 2002).

2.2 Summary thermodynamics

The driving force for the adsorption of a surfactant, intact droplet dispersal, mixing and formation of micelles and microemulsions is a minimisation of free energy (Adamson, 1990). The driving force is itself 'directed' by lower enthalpies (ΔH) and higher entropies (ΔS) (Hiemenz and Rajagopalan, 1997; Jones, 2002). When the entropy (disorder) is high, the free energy (ΔG) is low. Low free energies mean the process tends to be spontaneous. Adsorbed emulsifier is also subject to identical thermodynamic 'forces', and these can relate to ion binding/release and

Table 2.6 Thermodynamic influences and steric stabilisation mechanisms

Stabilisation	Result	Example
Enthalpic	↓ ΔH (reduces ΔG)	Mixing of palisade layers, water and ion release
Entropic	↑ ΔS (reduces ΔG)	Ordering of hydrophobic tail groups
Osmotic and restricted space	↓ ΔS & ↑ ΔH	Crowding of adsorbed layer, loss of conformation, influx of solvent

↓, reduction; ↑, increase.

permitted chain freedom, in a range of flocculated systems (see Table 2.6).

$$\Delta G = \Delta H - T\Delta S \qquad (2.2)$$

Binding and a range of processes occurring inside the oil droplet or micelle/microemulsion, or the sample in the case of an oily amalgam, are driven by thermodynamics (Equation 2.2). The 'hydrophobic effect', where self-assembly takes place, is driven initially by entropy (and its role in solubilisation). A similar end-point can also be achieved by increasing or reducing enthalpy with respect to entropy, so that $\Delta G \rightarrow 0$. Free energy is the primary driver behind the spontaneous establishment of a system (Jones, 2002). Compositional and local 'environmental' effects may make a further contribution (see Table 2.6).

2.2.1 Key influences and relationships

The nature of the oil and its density and viscosity (Hiemenz and Rajagopalan, 1997) are implicit in defying creaming and the melting point of the oil (in O/W emulsions), and its fluidity is fundamental to coalescence. The phase volume (φ) and the semisolid or fluidlike nature of the emulsion dispersion medium facilitate degrees of creaming due to a density variation. Increased diffusion, which relates to both thermal (Brownian) motion and particle size via Ostwald ripening, may then lead on to coalescence, which can mean that larger droplet sizes are determinants of 'relative stability' (Sinko, 2006; Mahato, 2007). The energetics of resistance to flocculation, known as Derjaguin–Landau–Verwey–Overbeek (DLVO) theory, were devised in the 1940s (see Section 3.2). The dispersed droplet-adsorbed-layer rheology (viscoelasticity) and its anticoalescence (Sarker *et al.*, 1995a,b, 1996, 1998a, 1999; Sarker, 2005a,b) and antiflocculation (Mahato, 2007) properties, given the close proximity of droplets and the increased incidence of collisions on warming and concentration in the creamed layer, have a large impact on avoidance of emulsion collapse.

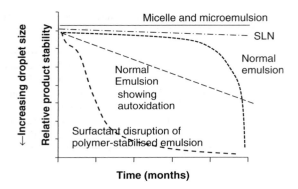

Figure 2.7   Indication of the relative stabilities for a range of emulsions. Emulsions are inherently unstable (metastable). 'SLN' refers to a solid lipid nanoparticle, where the emulsified oil is solid

Time and temperature (kinetics, rate of processes, solubility, rheology, diffusion and crystal form) are significant factors and agents in determining emulsion-quality and dosage-form PCQ. Additionally, the cycling of temperatures and pressures (e.g. air freight) can affect constitution, which may then modify polymorphism, leading to variations (and thus instability) in interfacial structure and mechanics. This is shown pictorially in Figure 2.7 and explains why solids lipid emulsions, such as SLNs, have a considerably better stability profile with respect to time than do liquid-based semisolids.

Stability is not attributable only to interfacial (emulsifier) and bulk structuring of both the dispersed phase and the dispersion medium but also to photocatalytic events, such as free radical generation as a result of autoxidation. Oxidation, rancidity and lipase action cause a downward spiral in the quality of the product and the evolution of noxious byproducts. Mechanical action and mixing, and their impact on particle size/distribution, permits droplet growth by Ostwald ripening (disproportionation). This will occur in partiuclar when combined with harsh pH conditions (as in oral emulsion delivery to the gastrointestinal (GI) tract or when passing through the stomach).

2.3 Interfacial tension and wetting

Adsorption of amphiphilic molecules (surfactants/emulsifier/polymers) at surfaces, with a subsequent reduction in surface free energy (surface/interfacial tension) and surface area expansion, drives spherical droplet formation (Sarker *et al.*, 1995a,b, 1996, 1999; Sarker and Wilde, 1999). Figure 2.8 shows that adsorption of an emulsifier on a surface permits wetting and spreading (Figure 2.8a) and a lowering of interfacial (surface) tension (Figure 2.8b). The extent of the lowering of interfacial tension depends on the emulsifier concentration and type. Polymeric emulsifiers are generally unable to achieve very low interfacial tensions (average

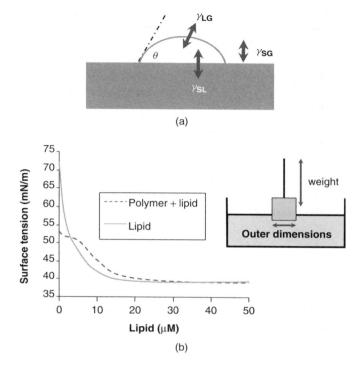

Figure 2.8 (a) Wetting of a surface by a sessile drop and three-line surface tension (γ) contributions to the spreading coefficient and to Young's equation for the contact angle (θ). Subscripts S, L and G refer to solid, liquid and gas, respectively. (b) Surface tension plot and schematic of an air/water (A/W) tensiometer (using a Wilhelmy plate tip), used to determine aggregation and critical micelle concentration or surface coverage. Oil-in-water (O/W) tensiometry uses a similar but marginally different Du Noüy ring apparatus tip

35–55 mN/m) when compared to LMW emulsifiers such as sodium dodecyl sulphate (~20 mN/m), which are able to form much tighter interfacial packing, better reducing surface free energy (interfacial tension).

Young's equation describes the line tension contributions to the wetting process, which is a combination of three interfacial tensions (Adamson, 1990). In pharmacy, wetting is required in order for diffusive processes and leaching or egress of drug to take place from DDS vehicles. Table 2.7 shows a range of HLB emulsifiers/surfactants and their customary roles in formulation. Implicit in this definition is their additional role in wetting or facilitated wetting, such as 'superspreading' phenomena (Rafai *et al.*, 2002; Castelletto *et al.*, 2003) and Marangoni processes (Marangoni, 1871). Adsorption is the first step in permitting a solvent (e.g. water) to wet a dry or oily surface.

Importantly, for any amphiphile (emulsifier), adsorption is about the minimisation of surface free energy. Surface (between gas and another phase) or interfacial (between any two distinct phases) tension (γ) is an expression of surface free

Table 2.7 Hydrophile–lipophile balance (HLB) and its use

Sample	Value (20°C)	Use
Hydrophile	15	Solubilisers, detergent
Hydrophile	12	O/W emulsifier
Lipophile	9	Wetting and spreading agent
Lipophile	6	W/O emulsifier
Lipophile	3	antifoams

W/O, water-in-oil.

energy under standard conditions. The two terms can be used interchangeably to some extent. The interfacial tension is the force that acts perpendicular to a line in the plane of the interface.

$$\gamma = Force/(2L) \tag{2.3}$$

Surface free energy (ΔG_s) is the work required to increase the area by $1\,m^2$ (under standard temperature, pressure), units mJ/m^2 ($J = Nm$); the work done is therefore γ (units mN/m). Emulsifiers differ in their ability to lower surface tension. LMW emulsifiers (Tween, Span, SDS, oleyl alcohol, etc.) can lower the interfacial tension to 20–25 mN/m, whereas polymers (e.g. poloxamers), proteins/peptides and hydrophobic gums, due to 'poor' space-filled interfacial packing, form an interactive surface layer, but one with higher (~50 mN/m) interfacial tension (Sarker et al., 1995a,b, 2005a; Collins et al., 2008; Concannon et al., 2010).

Some surfactants, such as those of the trisiloxane class, facilitate surface-structured wetting (Rafai et al., 2002), known as 'superspreading', and where two-character Janus particles are present they can also act as facilitators of wetting (Binks, 2002; Arditty et al., 2004; Concannon et al., 2010). This leads on to the notion of surface or interfacial spreading capability, a useful descriptor for topical products. This is defined in terms of the droplet spreading coefficient (S), which gives an idea of the ability of a liquid to spread on a surface and is defined as follows:

$$S = \gamma_A - (\gamma_B + \gamma_{A/B}) \tag{2.4}$$

where A represents solid substrate/against gas, B represents liquid/against gas and AB represents the line tension between solid and liquid. If the quantity is zero or positive, spreading is possible. Wetting is also defined by Young's equation, where the quantity, θ, represents the contact angle of a sessile drop on a surface:

$$Cos\,\theta = (\gamma_A - \gamma_{A/B})/\gamma_B \tag{2.5}$$

The composition of polymeric surfactants, their concentration and the extensivity of surface coverage and of the moieties hydrophilic and hydrophobic (lipophilic)

Table 2.8 Common linear block copolymer Pluronic (poloxamer) biocompatible emulsifiers

Pluronic® NF	Poloxamer	MW; % EO (#EO units)
L44	P124	2.1–2.4 kDa; 46 (12)
F61	P181	~1.8 kDa; 10
Lutrol Micro 68, F68 (Flocor)	P188; coating, dispersant, emulsifier (HLB 16)	7.7–9.5 kDa; 82 (80)
F87	P237	6.8–8.8 kDa; 72.4 (64)
F108	P338	12.7–17.4 kDa; 83 (141)
Lutrol Micro 127, F127	P407	9.8–14.6 kDa; 73 (101)

MW, molecular weight; EO, ethylene oxide; #, number of EO units.

portions also have a role to play in efficiency as a wetting agent and the capability for lowering interfacial tension (surface activity). This is shown in Table 2.8.

Emulsifier movement at the interface and rearrangement are rapid for a simple emulsifier. Nevertheless, 'ageing', or rearrangement of the interface materials, is necessary for the establishment of a definable interfacial composition. This is further complicated when the amphiphile is a large 'slow-diffusing' block copolymer surfactant or polymer/protein. Emulsifiers spread along the interface, lowering interfacial tension, covering the surface and giving the surface a protective film of increased surface viscosity (see Section 3.3). This influences product shelf life and can directly affect the attraction/repulsion between opposing particles (opposing coalescence).

Flow facilitated by wetting can consider Reynolds number, based on turbulence and driven by the rapidity of wetting:

$$\text{Reynolds number (Re)} = \rho u d / \eta \qquad (2.6)$$

where ρ is the density, u is the velocity, d is the size of the droplet and η is the bulk viscosity. Below 2040, flow is predictable and laminar rather than chaotic and turbulent. Also important are the spreading coefficient and, in lamellae separating droplets, the Laplace pressure, bulk viscosity and capillarity or capillary suction. On a molecular scale, the rugosity/roughness (Rafai *et al.*, 2002; Al-Hanbali *et al.*, 2006; Georgiev *et al.*, 2007) caused by the adsorbed molecule (e.g. pancake/mushroom or brush-configured adsorbed polymer) also determines fluid flow. For monolayers, as with bulk self-assembly, Traube's rule means that as the aliphatic portion of the surfactant/polymer grows, this reduces the concentration of association required for a micelle/planar micelle or biphasic separation to occur in the plane of the interface (Castelletto *et al.*, 2003). This is also the case for the bilayer of a liposome or the monolayer of an emulgent on an emulsion droplet, and can act as the phase for better emulsification for high log *P* drugs.

Table 2.9 Common nonlinear block copolymer Tetronic (poloxamine) biocompatible emulsifiers

Tetronic®	Polymeric micelle	Use
150 R1 (large PPO)	Y	Large hydrophobic core
304	Y	Lipofection; gene delivery
701	Y	Surfactant (3.6 Da)
901	Y	Surfactant
908	Y	Emulsifier

Table 2.9 indicates how block copolymer emulsifiers can self-assemble to produce micellar aggregates with an 'oil-like' or hydrophobic core when formed in aqueous media. Block copolymer emulsifiers are used widely for solubilisation of apolar drugs.

The surface concentration of polymer or simple emulsifier is explained by the Gibbs equation (Equation 2.7). Concentration in the bulk phase and interspecies competition drive the packing and form of the adsorbed layer of emulsifier. Other parameters which influence interfacial composition include bulk viscosity, interfacial rearrangement and competitor molecule desorption.

$$\text{surface excess concentration } (\Gamma) \quad = (n_i^{tot} - n_i^A - n_i^B)/\text{area} \qquad (2.7)$$

where n is number of moles of phases A and B.

2.3.1 Emulsifier types

The most commonly used emulsifiers are nonionic varieties, such as polysorbates (Tween 20–85), sorbitan esters (Span 20–85), polyethers (e.g. Brij35) and block copolymers, including PEG or the well-used poloxamers (Pluronics) found ubiquitously in pharmacy products (Sarker, 2008, 2012a). Medical use is driven by their robust chemical nature and their ability to resist small changes in environmental temperature, pH and ionic strength. Table 2.10 shows the types, properties and uses of these functionalising emulgents.

The word 'surfactant', used interchangeably with 'emulsifier' in this text, derives from the term 'SURFace ACTive AgeNT' (coined by Antara Products ca. 1950). Whatever the form, geometry and HLB, they possess the ability to lower interfacial tension (γ). Surfactants and emulsifiers (surfactants used for emulsification) have a Krafft point (T_k) temperature, which refers to nonionic emulsifiers for micellar systems, in which there is a marked increase in solubility above this temperature. Below the Krafft point temperature, increasing concentration only leads to precipitation. The cloud point (PIT) is the temperature which separates surfactant (emulsifier) precipitation from gel formation of an initial cloudy

Table 2.10 The four basic forms of emulsifier used in pharmacy products

Type	Example
Anionic	Fatty acids, sodium dodecyl sulphate (SDS)
Cationic	Cetrimide, C_{16} (many toxic), phosphatidylethanolamine (PE)
Non-ionic	Polysorbates (Tweens), Brij's, sorbitan esters (Spans), cetomacrogol (POE ethers), Pluronics, hydroxypropyl (methyl cellulose) (HPMC), solid particles
Ampholytic	Lecithin, proteins (lung surfactants), membrane components

solution, and this is important in terms of optimal droplet coverage (just as Krafft temperature is) in emulsification. PEG polymer, also known commercially as Carbowax/Macrogol (general formula C_xE_y, describing the number of carbons in the hydrophobic part and hydrophilic ethylene oxide portions, respectively), is used when MW<20 kDa; polyethylene oxide (PEO) when MW>20 kDa; and polyoxyethylene (POE) for general use and of any molecular mass. These molecules are used ubiquitously in pharmacy and can have differing geometries, including branched, star and comb types. The molecule is highly favoured because of its 'simple' structure, and once in the blood PEG shows slow clearance (e.g. removal by the kidneys), meaning longer-acting capability. The most valuable PEG size appears to be 2–5 kDa, because above this peri- or transcellular permeation is inhibited and circulatory (blood) clearance is facilitated.

Polymeric emulsifiers fall into two categories:

- Homopolymers, e.g. PEG.

- Heteropolymers, which consist of a repeating-unit block copolymer, e.g. Pluronics (F68, F108), Tetronics (901, 908) and have the form:

  - XYXYXYXYXYXY (alternating);

  - XXXYYYXXXYYY (blocks);

  - XYXXYXXXYXXY (random).

where X and Y are distinct monomer units. Other surface active polymers (Sarker and Wilde, 1999) include hydroxymethyl-, ethyl- and propyl-methyl-cellulose (Sarker et al., 1999), chitosans, gums and PEGylated lipids (PEG-lipids). Simple emulsifiers are usually modelled on the 'lollipop or tadpole' format: a circular dense polar headgroup and a linear single or multiple alkyl tail (see Figure 2.9).

An important description of an emulsifier format is the HLB, coined by Griffin in 1954. This is the ratio of hydrophilic to hydrophobic parts, with low values relating to antifoaming oils and higher values referring to the solubilisers that

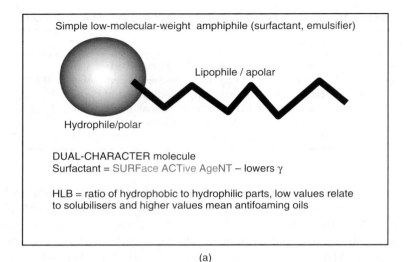

(a)

(b)

Figure 2.9 Schematic representation of (a) a simple low-molecular-weight (LMW) emulsifier (typically <1.5 kDa) and (b) a polymeric emulsifier (typically 5–40 kDa). The terms 'emulsifier', 'surfactant' and 'amphiphile' are interchangeable

define the emulsifier use. The hydrophobic portion of the molecule has an assigned value (-CH$_2$-, -CH$_3$ and = CH- is 0.48), which can be used with a knowledge of the molecule to calculate an overall value. This is also the case for assigned polar groups (-COO- is 19, -OH is 2, -C = O is 2.5 and a sugar alcohol group is 0.5) and for oils themselves (paraffin ~9, silicone ~11, vegetable oil ~8). Generally, this can be described as:

$$HLB = 20\,[1 - (S/A)] \tag{2.8}$$

where S is the saponification value and A is the acid value. For Tween 20, this would be $20(1 - (46/276)) = 16.7$.

For polyoxyethylene ethers it is:

$$HLB = ((E + P)/5) \qquad (2.9)$$

where P is the percentage by weight of polyhydric alcohol (glycerol or sorbitol) and E is the percentage ethylene oxide or (number $\times$ molecular weight of the fraction)/molecular weight of the emulsifier) multiplied by 100. Where oxyethylene surfactants are used:

$$HLB = (E/5) \qquad (2.10)$$

where E is the percentage of ethylene oxide units per molecule.

HLB matching of surfactant mixtures can be used to best formulate the size of emulsion droplets. In industry, this method is used because the composition is thus not entirely linked to a 'pure' single emulsifier.

$$HLB_{mix} = (f \times HLB_A) + [(1 - f) \times HLB_B)] \qquad (2.11)$$

where A and B are two emulsifiers and f is the fraction (e.g. percentage of the total) of a particular HLB emulsifier. A similar (accordingly modified) arithmetic formula would be used for three or more emulsifier blends. Single or multiple emulsions can be fabricated by using emulsifiers of different shapes and HLB values, such as Tween 20 (HLB, 17) and Span 80 (HLB, 4), in separate emulsifications (see Figure 2.10).

Surface properties become crucial in emulsification processes. Once adsorbed at the interface, the polymeric emulsifier shape depends on the local environment and the polymer size, hydratability, shape, flexibility and molecular weight. Soft condensed matter (Jones, 2002), also referred to as soft matter (including polymer dispersions and nanogel beads), self-organises into microscopic 'devices' and thus into mesoscopic physical structures that are much larger than the atomic scale yet smaller than the eye can see. This structuring determines the macroscopic behaviour of the material and, in the case of emulsions, rheological texture, transparency/opacity and shelf life. There are a number of soft matter types:

- Surfactant (emulsifier) structures, e.g. micelles, LCs, foams.

- Lipidic structures, e.g. liposomes/vesicles, cell membranes, composite materials.

- Polymer structures, e.g. gels, complex fluids, multielement composite materials.

Consequently, soft materials are very important (crucial) in a wide range of technological applications habitually encountered by pharmacists and biomedical scientists (Sarker, 2006b, 2008, 2010, 2012a; Collins *et al.*, 2008), including:

- IV and nasal fluids, pastes and topical forms.

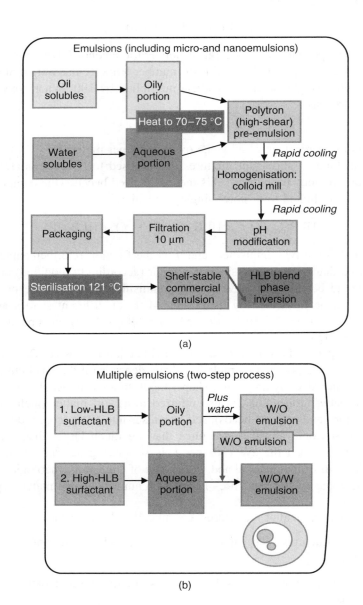

(a)

(b)

Figure 2.10   (a) Route to general fabrication of a coarse pharmaceutical emulsion. (b) Fabrication of a coarse pharmaceutical multiple emulsion (and emulsion within an emulsion). Notably, multiple emulsions require two distinct HLB emulsifiers

- Emulsions (as used in the above and for all pharmaceutical creams).

- Cosmetology products (cosmeceuticals) and drug product additives (e.g. flavour and colour encapsulation).

Micelles are aggregates (colloids) in which monomeric amphiphiles undergo association in a process called micellisation, which was first studied by McBain in 1913. This self-association is due to the hydrophobic effect (see Equation 2.2). The concentration at which this occurs is known as the critical micelle concentration (CMC), or sometimes the critical aggregation concentration (CAC) for surface active polymers. The surfactant charge, size, shape and solvent environment define the CMC (see Figure 1.2). Micelles are usually found in a normal (common) or reverse (inverted) format (about 5–10 nm), as illustrated in Table 2.3. Most micelles are composed of 100 or so surfactant molecules. The CMC for a particular emulsifier can be determined by physicochemical changes in surface tension, absorbance, density, conductivity, osmotic pressure and so on. The most commonly used is the surface tension (Sarker et al., 1995a,b, 1996, 1999).

Notably, as a general rule the CMC:

- Decreases as the apolar part (alkyl chain) of the molecule gets bigger (Traube's rule).

- Increases as the polar part of the molecule increases.

- Shows a decrease for charged surfactants plus ions, as this causes the micelles to 'swell'.

Micelles form through a minimisation of free energy (ΔG). The change in water structure around the hydrophobic part of the surfactant is associated with a decrease in entropy (ΔS; this is the most influential contribution to overall thermodynamics: see Equation 2.2). Micellisation thus causes a favourable increase in entropy (due to an increase in disorder) and is therefore spontaneous. Polymerised micelles can be created by block copolymers (Pluronics) and PEGylated surfactants/lipids (see Tables 2.8 and 2.9), which are used for tetanus toxoid encapsulation and many other pharmacy roles.

In addition to their traditional drug delivery roles, liposomes and vesicles are finding numerous applications in encapsulation of flavours and as cryoprotectives for gene therapy agents (Richards et al., 2004; Hayes et al., 2006). The potency of the liposome as an 'emulsification' system for drugs has been and continues to be widely exploited for dermal and parenteral DDSs. Paul Ehrlich initially conceived of targeted delivery in 1909, when he envisaged a drug delivery mechanism that would be specific to a particular diseased cell. This was followed by a sea-change moment when Alec Bangham (Babraham Institute) developed the liposome (a.k.a. Bangasome) from an interest in cell membranes in 1961 (Horne et al., 1963;

Bangham and Horne, 1964). Importantly, the LC format of the lipid used to make the bilayer (leaflet) and various phase transitions can be used to facilitate better drug delivery (Sarker, 2009a,b, 2012a). The phase transition of liposome lipids changes with molecular weight (see Section 2.1 and Figure 2.4). These transitions or changes in state can include:

- DMPC, di-myristoylphosphatidylcholine (lecithin; phosphatidyl choline; PC), $T_m = 24°C$;
- DPPC, di-palmitoyl PC, $T_m = 41°C$;
- DSPC, di-stearoyl PC, $T_m = 55°C$;
- DOPC, di-octoyl PC, $T_m = -20°C$;
- 50 : 50 DMPC + DPPC, $T_m = 37°C$ (core body temperature).

2.3.2 Pickering emulsions

Pickering emulsions had not been discussed until recently (for drug delivery purposes), but this is now changing, with, for example, drug encapsulation in crystalline surface materials (Chen *et al.*, 2006). This makes use of an emulsifying agent, which may include finely divided solids, such as Janus particles (twin-faced single particles, with each face either lipophilic or hydrophilic). Pickering emulsions (Ramsden, 1903; Pickering, 1907; Aveyard *et al.*, 2003; Concannon *et al.*, 2010) are droplets stabilised by solid or semisolid particles (e.g. micelles and liposomes). They offer great opportunities for housing drug. Currently, most examples in research use silica or latex particles for Pickering stabilisation. The wetting contact angle (Binks, 2002; Aveyard *et al.*, 2003; Arditty *et al.*, 2004) has a large role to play in surface adsorption processes. The resilience of the particle at the interface and the capability for interlocking of particles, since fusing of particles forms a permeable shell, provide interesting opportunities. In light of this, a new form of particle (a 'colloidosome') has been postulated for drug encapsulation (Dinsmore *et al.*, 2002). A three-tier DDS (drug in particle, in oil droplet and at surface adsorbed layer) based on soft matter in the form of Pickering emulsions has also been proposed (Concannon *et al.*, 2010).

2.4 Shear and size reduction

Fundamental to both coarse and nanoemulsions (not micro- or micellar colloids) is the employment of shear and size-reduction equipment at shear stresses of MPa and at frequencies of agitation of kHz magnitude. Hot preparation (pasteurisation at 75°C) of the emulsion ensures a low microbiological bioburden and thus a

Table 2.11 Commercial and pilot-scale high-shear mixers and mills used to create coarse and fine oil droplet dispersions

Example of mixer/blender or mill	Mid-range particle size (diameter)	Primary use in emulsification	Types of sample
Silverson high-shear blade device	50 μm	Premixing	O/W emulsions, phase volume <0.7
Polytron PT-MR 3000 blender and antifoam spindle (PT-DA3012:2 TS) high-shear blade device	50 μm	Premixing	O/W emulsions, phase volume <0.7
Avestin mill (~100 MPa shear)	0.5–2.0 μm	Fine emulsion	O/W emulsions, phase volume <0.2
Hielscher ultrasound emulsificator and industrial ultrasonic processors	0.3–1.0 μm	Fine emulsion	O/W emulsions, phase volume <0.7
LabPlant (or PMT) jet mill and French press	0.3–1.0 μm	Fine emulsion	O/W emulsions, phase volume <0.5

O/W, oil-in-water.

better shelf life (Sarker, 2008). However, the enhanced fluidity of the vehicle under a high shear field means the dispersion phase (normally oil in pharmacy) is stretched and sliced to form ever smaller droplets. Use of a high-shear mixer, colloid mill, French press, Silverson-type blender, ultrasonic probe and impellar are all utilised routinely (Sarker, 2006b; see Table 2.11). For effective, uniform and better-quality monodispersion (or near monodispersion), most emulsions require a premixing step, in which fluid flow and turbulence are considered and modelled carefully. High energy input is needed to permit homogenisation or to produce sufficiently small droplet sizes for the coarse emulsion (Sarker *et al.*, 1999).

2.5 Raw materials

Raw materials are covered throughout this book, but it is important to describe the formulation aids essential to most forms of pharmacy emulsion product. These include the:

- Emulsifier (various types), lipid, polymer, thickener, gum (stabiliser, viscosifier).

- Oil (fat, should be liquid at the point of homogenisation), with correct measures to prevent autoxidation.

- Buffered, purified water.

- Drug or active pharmaceutical ingredient (API; active).

Purity, form and predictability of composition and physicochemical properties are essential to ensuring product PCQ (Sarker, 2008). The additives most frequently used to confer an extension of shelf life in pharmaceutical emulsions include (Russell, 1991; Ansel *et al.*, 1999; Lupo, 2001; Alayoubi *et al.*, 2013):

- Antioxidants to counter free radical formation and rancidity, e.g. propyl gallate, ascorbyl palmitate (a form of vitamin C), tocopherol (vitamin E), BHA, butylated hydroxytoluene (BHT), quercetin (Di Mattia *et al.*, 2010) and other flavonoids.

- Chelation of metal ion prooxidant species, e.g. sodium edetate (ethylenediamine tetracetic acid, EDTA), glycine.

- Buffering agents in the aqueous phase, e.g. acetates, phosphates, citrates, fumarates and their acid forms; amino acids; and alkalinising agents, e.g. sodium hydroxide, triethanolamine.

- Chemical preservatives, e.g. chlorocresol, benzoic acid/benzoates (parabens), sorbates, salicylic acid, p-hydroxycinnamic acid.

- Isotonicity agents, particularly for IV formulations, e.g. sodium/potassium/calcium chlorides, dextrose.

- Humectants, e.g. glycerine, sorbitol, propylene glycol.

3

Stability, metastability and instability

Dispersions are metastable entities; that is, they are thermodynamically unstable despite numerous attempts (see Sections 1.3, 2.2, 3.2.1, 3.3.3 and 15.2) to describe 'stable emulsions'. It is groundless to refer to a coarse emulsion (or pharmaceutical product) as being 'stable'. Figure 3.1 successfully captures the diversity and interconnection between processes that compromise or reduce the 'relative stability' of the finished product. The key processes defining the emulsion shelf life are flocculation, creaming, coalescence and Ostwald ripening. The first three are augmented when the concentration of droplets (phase volume, φ) is higher; that is, when $\varphi > 0.5$ to $\varphi > 0.74$ (loose packing of droplets up to $\sim 0.5\varphi$).

Coalescence and temperature are related via the Arrhenius equation:

$$C = k \exp^{-(E/RT)} \qquad (3.1)$$

where C is the extent of coalescence, k is the collision constant, E is the potential energy barrier (in oil-in-water (O/W) emulsions, this is a function of electric, steric or van der Waals potential and parameters; see Section 3.2.1).

Needless to say, warming the sample increases the kinetic energy, meaning more fatal collisions of the emulsion droplets. This coalescence is compounded according to Equation 2.1 as the dispersed oil fluidity increases with temperature. Similar perturbation can occur in both the stomach (see Section 12.4.3) at low pH and the gut under the influence of dietary lipids and proteins, which compete for interfacial coverage (see Sections 3.3.1 and 3.3.2). This leads to coalescence, and for this reason formulations of the coarse emulsion type are not used routinely for gastrointestinal (GI) tract drug delivery. Flocculation by depletion mechanisms (insufficient emulsifier used) and the 'osmotic' driving force based on mutual low concentration/coverage forces particles together (coalescence). Where droplets are conjoined by emulsifier, this is referred to as 'bridging flocculation'. The mechanisms of flocculation are illustrated phenomenologically in Figure 3.2.

Another important determinant of product PCQ (purity, consistency, quality) is creaming (covered in Section 3.1). Its importance lies in the way in which

Pharmaceutical Emulsions: A Drug Developer's Toolbag, First Edition. Dipak K. Sarker.
© 2013 John Wiley & Sons, Ltd. Published 2013 by John Wiley & Sons, Ltd.

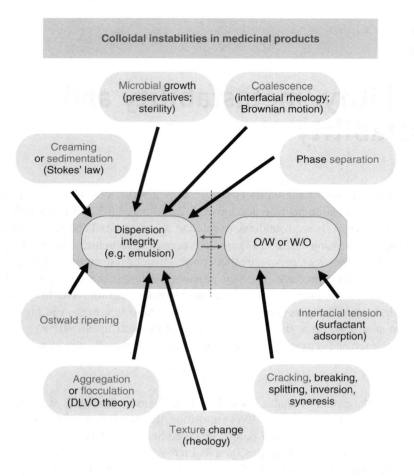

Figure 3.1 Emulsion 'stability' factors and conditions driving the physical degeneration of the emulsified form. Some factors are interlinked, e.g. flocculation (conjoining) can promote creaming (floatation), which may then under conditions of particle concentration and crowding promote coalescence (droplet rupture)

it can lead on to other processes. The consequence of repeated coalescence can be emulsion breaking (syneresis, cracking, splitting, etc.) or phase inversion of the emulsion, e.g. O/W → W/O. Ostwald ripening (disproportionation of droplets) occurs when the very high pressure inside smaller droplets forces the dispersed phase to 'dissolve' in the continuous phase and to be reconstituted in larger (low-pressure dispersed phase) particles. The process is also true for dispersed solids such as solid lipid nanoparticles (SLNs). The net effect of this process is that only very large droplets remain, increasing the overall catastrophic result of having droplets in close proximity (creamed layer) and of the occasional thin film rupture.

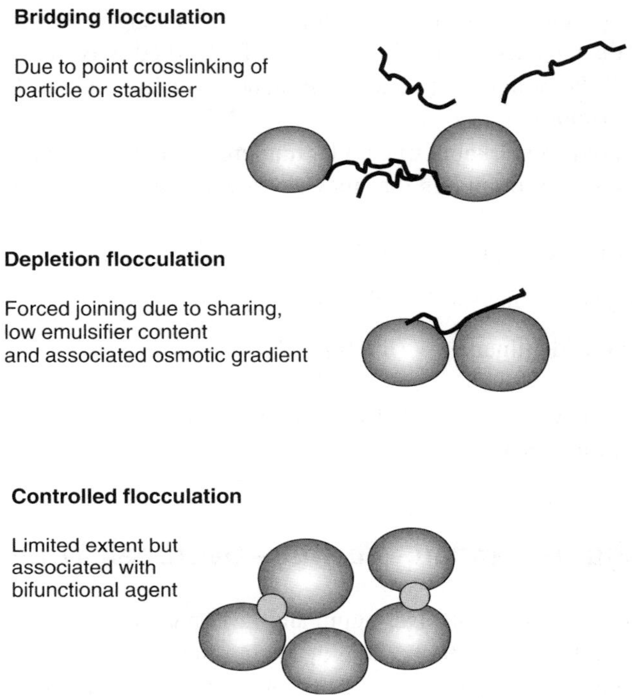

Bridging flocculation

Due to point crosslinking of
particle or stabiliser

Depletion flocculation

Forced joining due to sharing,
low emulsifier content
and associated osmotic gradient

Controlled flocculation

Limited extent but
associated with
bifunctional agent

Figure 3.2 Different types of droplet flocculation and joining and their root causes

3.1 Stokes' law

Stokes' law concerns droplet creaming and sedimentation. The velocity of the
process is governed by the following relationship:

$$\upsilon = \frac{2R_g^2(\Delta\rho)g}{9\eta_o} \tag{3.2}$$

where the density difference is $\Delta\rho = \rho - \rho_0$ This can be accelerated in a cen-
trifuge (perikinetic creaming to test the shelf life in an accelerated form), in which
case g is replaced by $(\omega^2 X)$, the angular velocity and distance. In Equation 3.2,
R_g is the radius of gyration (radius) of the droplet, ρ is the density of the particle,
ρ_0 is the density of the solvent, η_0 is the viscosity of the medium and g is the
acceleration due to gravity (9.81 m/s^2). The gravitational constant can be ignored
under standard conditions. Put quite simply, density difference, particle size and
viscosity are determinants of the rate of creaming. For O/W dispersions, one
would normally expect to see creaming, as the density of most oils is significantly

lower ($0.7-0.9 \, g/cm^3$) than that of water (see Table 2.4). The inverse might be true for W/O dispersions, although the effect of adsorbed polymer or emulsifier in modification of the bulk density and its influence on creaming has been discussed (Sarker *et al.*, 1999a).

The theory holds remarkably true under numerous conditions and circumstances, but it makes some assumptions that may not always be relied upon:

- Spherical particles present (likely).

- No aggregation (may not be likely).

- Streamline or laminar flow (likely).

For small particles ($<2-5 \, \mu m$), Brownian motion counteracts sedimentation or creaming to aid longer-term dispersal.

3.2 Derjaguin – Landau – Verwey – Overbeek (DLVO) theory

Four independent scientists, B. Derjaguin and L. Landau (ca. 1941) and E. Verwey and J. T. G. Overbeek (ca. 1948) came up with a theory of colloid and dispersion stability (Hiemenz and Rajagopalan, 1997), which was unified and called DLVO theory. DLVO theory (Equation 3.3) describes coagulation (and related processes) and aggregation of droplets or dispersed particles. This results from permanent particle–particle contacts, producing typically large aggregates. Irreversible coagulation occurs when ζ-potential or surface electrical charge is too low or permits interparticle mediation by local counterions (gegenions). Reversible coagulation is seen and controlled by the magnitude of secondary minimum in the potential energy distance plot (see Figure 3.3), and this relates to larger particles ($>1 \, \mu m$) such as coarse emulsions. The theory can be summarised simply as stating:

$$V_T = V_A + V_R + V_S \qquad (3.3)$$

where V_T is the total interaction energy, V_A is the attraction potential of two entities, V_R is the repulsion potential of two entities and V_S is the steric (attraction or repulsion) potential of two entities. The magnitude of attractive (V_A), repulsive (V_R) and steric (V_S) 'forces' and their interplay governs whether sample flocculation and coagulation occur. The theory can be applied universally across a range of media, covering polymer micelles, liposomes and microemulsions, and then to solid surfaces, to droplets, to solid dispersed particles, to bubbles, to thin liquid films (TLFs) and to coarse emulsion droplets (Castelletto *et al.*, 2003).

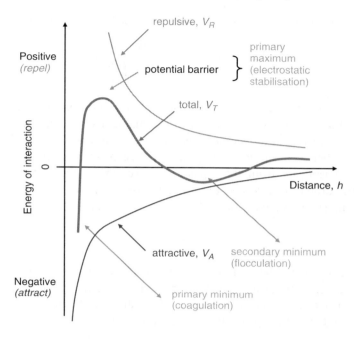

Figure 3.3   Representation of the potential energy of interaction diagram according to the Derjaguin–Landau–Verwey–Overbeek (DLVO) combined theories. Principal elements of interest are the repulsive (V_R) energy contribution, attractive (V_A) energy contribution and a steric (V_s) energy contribution, which may pertain and be additional to either of the former energies. The summation of all these individual contributions gives the total energy of interaction, with a notable (and typical) primary minimum, primary maximum and secondary minimum at various distances between dispersed particles

3.2.1 DLVO energy diagram

Figure 3.3 describes the relative contributions of three energy terms and the summation of these to produce an interaction energy (V_T)–distance (h) curve with built-in descriptions of the estimations of potential for compositional change due to overlap of electric double layers from Gouy–Chapman theory (V_R) and electronic polarisation within atoms from London dispersion forces theory (i.e. nonpolar molecule polarisation), a form of van der Waals forces theory (V_A).

Studies undertaken on the TLF that separates two oil droplets (or two air cells, for optical convenience) have been used to derive and estimate the 'disjoining pressure' (a form of representation of the features of DLVO theory and theoretical parameters; Castelletto *et al.*, 2003) that describe molecular interactions, such as the Hamaker constant (A). The repulsive potential can be calculated as follows:

$$V_R = 2\pi\varepsilon\varepsilon_0 a\psi_0^2 \times \frac{\exp(-kh)}{1 + \left(\dfrac{h}{2a}\right)} \qquad (3.4)$$

where π is a geometric parameter, ε and ε_0 are permittivities of the media, a is the radius, ψ_0 is the surface potential (analogous to ζ-potential), κ is the reciprocal Debye length (concerning the ion sheath around the particle) and h is the distance (between entities). For small surface potentials (charges), the term $\left(1 + \left(\frac{h}{2a}\right)\right)$ disappears from the equation. In summary, the distances between droplets, the droplet charge and the droplet/particle size mostly have an impact on repulsion. The attractive potential can then be estimated as follows:

$$V_a = \frac{-Aa}{12h} \qquad (3.5)$$

The attractive potential is mostly determined by the species polarisation or interaction energy (Hamaker constant) and particle size. The steric contribution (V_s) may fall into either or both 'camps', depending on the nature of the surface coverage or the chemical properties of the emulsifier.

Debye length and decay

Charge decay moving away from the particle surface and the succession of ion and counterion sheaths can be described by the Debye length (κ). The first layer of counterions after the actual particle is referred to as the Stern layer. Consequently, $1/\kappa$ relates to the Stern layer end and the decay length.

Gouy–Chapman theory

Gouy–Chapman theory determines the extent of charge screening and charge-based particle repulsion. It goes on to describe the electrostatic remnants of excess surface charge in the form of multiple layers of gegenions extending away from the particle surface.

Hamaker modelling and constant

Hamaker modelling and the constant itself describe the relative affinity of similar and dissimilar particles, atoms or molecules for each other at a fundamental level, based on a mapping of all possible permutations and combinations in a spatial arrangement within the system. The value can be defined by:

$$A = 12H/r \qquad (3.6)$$

where H is the summation of pairwise molecular interactions and r is the spacing distance. The modelling is complex and out of the scope of this textbook, but simulations and various models have been used to describe the magnitude of the energy of interaction. Dielectric constants feature strongly in determining the magnitude of the Hamaker constant.

London (van der Waals) dispersion forces

Also referred to as association or dispersion forces, London dispersion forces are the result principally of electronic polarisation, depolarisation and induced polarisation of the electrons associated with the atoms or molecules of the species constituting the dispersed particle or oil droplet. The summation of these energy contributions gives the particle a defined and specific value (see Equation 3.6).

Steric 'stability' factors: palisade layer

When an emulsifier with a large hydrophilic group (generally of the polymer type) adsorbs on the surface of an oil droplet, the palisade layer of polar emulsifier extending away from the particle surface gives the particle particular mechanical and chemical properties and defines the likelihood of coagulation and flocculation (see Figure 3.3). Coalescence, Ostwald ripening and entanglement of the palisade layer components and flocculation can occur, and may not depend on the form of the emulsifier. Ionic charge relating to surfactant pK_a for charged emulsifiers, such as fatty acids or lecithin, may also be significant in stabilisation. Polymer conformation in the form of surface 'brush', intermediary states or 'mushroom' and the lower critical solution temperature (LCST), which can change form, much like polymeric micelles of block copolymers, all have a role to play in both maintaining the integrity of the particle from its neighbours and, due to DLVO theory, avoiding opsonisation in the case of colloidal plasma proteins in the blood for injectable oil droplets (Al-Hanbali et al., 2006).

In a nonflocculated system (e.g. deflocculated), individual particles remain as discrete entities and the palisade layer is neither entangled nor breached. However, it may be advisable to allow some degree of controlled flocculation based on surface charge, as influenced by electrolyte concentration (e.g. bifunctional ions), and use this to impart increased bulk viscosity. Flocs are groups of weakly bonded particles that can impact on drug product efficacy and suitability. Control of rheological properties is important (crucial) to maintaining the way in which the product is presented pharmaceutically.

Restabilisation (ΔG, ΔH and ΔS) and interfacial structure In a dispersion, given that the particles are covered with a layer of emulsifier, it is possible to allow a form of stabilisation (via ΔG, ΔH and ΔS) of individual droplets, based on lowering of the free energy via various mechanisms and interfacial structure restrictions indicated in Table 2.6. The table explains the effect of modification of the particle surface and adsorbed polymer/emulsifier in terms of favourable form, conditions and restrictions to form, and the impact this has on restabilisation.

High ionic strength effects, destabilisation and flocculation   Where formulations encounter higher ionic strengths (μ), destabilisation and flocculation are commonplace. Depletion and bridging flocculation can then be favoured, leading to creaming and product splitting. Screening of charge and removal or addition of ionic or electrostatic repulsion effects may lead to flocculation, meaning a clustering of droplets. Problems also occur with pharmaceutical samples for injection (intravenous (IV) products) in 'toying' with isotonicity, using NaCl or $CaCl_2$ additions, which modify surface charge but are needed for therapeutic tolerance.

3.3 Interfacial rheology

Interfacial viscoelasticity serves two mechanisms of droplet stabilisation. In the first instance, it provides a barrier to coalescence (providing a water cushion), via pendant tails, loops and trains, intermolecular interaction and potential crosslinking. Water binding (in 'stealthification' of medicinal nanoparticles) is central to particle stability and tolerance. In the second instance, the emulsifier/polymer shows a 'continuity' within the insolubilised surface layer of amphiphile, thus providing a pure mechanical barrier to prevent coalescence. Both mechanisms have been widely investigated (Chen *et al.*, 1993; Sarker *et al.*, 1995a,b, 1996, 1998a,b, 1999a,b; Wüstneck *et al.*, 1996; Sarker and Wilde, 1999; Sarker, 2005b; Collins *et al.*, 2008; Woodward *et al.*, 2009) and have led to the phenomenological surface model shown in Figure 3.4. The methodology behind these discoveries is discussed in Section 3.3.1.

Close examination of the surface of a droplet (or equivalent model) indicates the possibility for monolayer versus multilayer form (Chen *et al.*, 1993; Sarker *et al.*, 1998a). This has been routinely determined by specialised surface rheological techniques and forms of spectroscopy or microscopy such as dilational/shear/bulk or atomic force microscopy (AFM) rheology, the fluorescence recovery after photobleaching (FRAP) photophysics technique (Yguerabide *et al.*, 1982) and interferometry.

When a polymer surfactant adsorbs on the surface of a droplet or any other medium, it establishes a comprehensive network of intermolecular interactions (Sarker *et al.*, 1998a,b), described adequately by Equation 3.7. Entanglement and interaction are decisive in determining cohesiveness and elasticity (Equation 3.8) and molecular mobility (Equation 3.9). Equations 3.8 and 3.9 are determined experimentally (and described in Chapter 14). The parameters E_d, E (a composite of elasticity and viscosity) and D describe the surface dilational elasticity, elastic modulus and surface diffusion coefficient, respectively. The quantity D can be used to chart surface viscoelasticity and elasticity or rigidity spectrophotometrically, as it describes the 'compartmentalising effect' of a surface polymer or an emulsion on the restriction in diffusion of a fluorescent probe or drug analogue molecule (Sarker et al., 1995a; Sarker, 2005b).

$$G_n = \rho_c \left(M/M_{ent} \right) \cdot \left[(k_B T) + K_c \cdot \rho_c^2 \right] \tag{3.7}$$

where G_n is the elasticity of the network, M is the molecular weight, M_{ent} is the entanglement in solvent, c is a critical degree of entanglement, ρ is the density of the connections, k_B is the Boltzmann kinetic constant, T is the temperature and K is a critical constant that must be obtained to facilitate interconnection.

The surface dilational elasticity (and related quantity E) is determined from:

$$E_d = \Delta\gamma/(\Delta \ln A) \tag{3.8}$$

where γ is the interfacial tension and A is the surface area. Similarly to dilational rheology, but using alternative methodology, the lateral diffusion coefficient (D), in units of cm^2/s, a measure of surface rigidity, is determined from the fluorescence recovery time (t_D) and the radius of the illuminated laser spot (ω) using the follow the process:

$$D = \omega^2/t_D \tag{3.9}$$

3.3.1 Gibbs–Marangoni hypotheses (viscosity or elastic modulus)

Generally referred to as the Marangoni (Marangoni, 1871) or the Gibbs–Marangoni hypothesis, this relates emulsifier surface mobility (Figure 3.4) to interfacial viscosity or elasticity. The theory generally concerns emulsifiers that show significantly hydrophilicity and so have a marked ability to associate with water. Low-molecular-weight (LMW) emulsifiers (surfactants) have rapid diffusion (higher D values); the hydrophilicity of these molecules has the effect of dragging solvent along with the emulsifier as it moves laterally within the plane of the interface, ultimately permitting a recovery of tension gradients and avoidance of TLF rupture, thereby maintaining the separation of oil droplets form one another (Sarker et al., 1995a,b, 1996, 1998a,b, 1999; Sarker, 2005b). Such gradients frequently arise due to thermal or mechanical perturbation of the emulsions as part of creaming and compaction or warming. Maintaining TLF thickness between oil droplets is essential to preventing droplet fusion.

3.3.2 Polymers at surfaces: Wilde–Sarker–Clark polymer hypotheses

Understanding of complex interfacial mechanics arises from a combination of hypotheses from the researchers Peter Wilde, Dipak Sarker and David Clark (Chen et al., 1993; Wilde and Clark, 1993; Dickinson and Hong, 1994; Sarker et al., 1995a,b, 1996, 1998a,b, 1999; Sarker and Wilde, 1999; Sarker, 2005b; Woodward

Thin liquid films – microscopy images

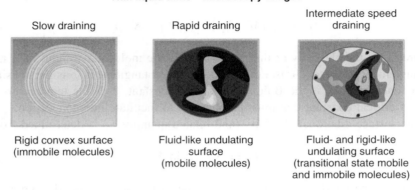

Slow draining	Rapid draining	Intermediate speed draining
Rigid convex surface (immobile molecules)	Fluid-like undulating surface (mobile molecules)	Fluid- and rigid-like undulating surface (transitional state mobile and immobile molecules)

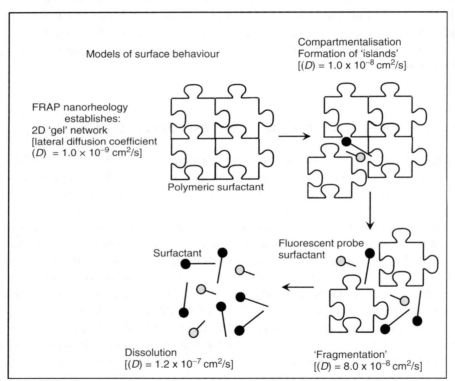

Models of surface behaviour

Compartmentalisation Formation of 'islands' $[(D) = 1.0 \times 10^{-8}\,cm^2/s]$

FRAP nanorheology establishes: 2D 'gel' network [lateral diffusion coefficient $(D) = 1.0 \times 10^{-9}\,cm^2/s$]

Polymeric surfactant

Surfactant

Fluorescent probe surfactant

Dissolution $[(D) = 1.2 \times 10^{-7}\,cm^2/s]$

'Fragmentation' $[(D) = 8.0 \times 10^{-8}\,cm^2/s]$

Figure 3.4 Phenomenological model of the surface structure of a droplet/particle covered with either a low-molecular-weight (LMW) surfactant, a polymer/protein or a mixture of both. The hypothesis is drawn together by a combination of surface diffusion (fluorescence recovery after photobleaching, FRAP), microscope observations from thin-film drainage and interferometric measurements of a thin liquid film (TLF). Refer to Sarker *et al.* (1995a,b)

et al., 2009). Such theories are modified to the particular form of dispersion. They summarise the result of an effective insolubility of polymer (e.g. protein) adsorbed in the plane of the interface. The adsorbed molecular material interacts laterally (Figure 3.4), increasing both elasticity and network continuity (Al-Hanbali *et al.*, 2006), and in the form of surface insolubilised monolayers or even multilayers (Clark *et al.*, 1991a,b; Courthaudon *et al.*, 1991; Wilde and Clark, 1993; Dickinson and Hong, 1994). This has been likened to a two-dimensional gel (Sarker, 2005b). The lateral diffusion for materials at the surface can be of the order of: monolayer of emulsifier $D = 1.2 \times 10^{-7}\,\text{cm}^2/\text{s}$; monolayer of polymer $D = 1 \times 10^{-8}\,\text{cm}^2/\text{s}$; biphasic mix of polymer and emulsifier $D = 8.0 \times 10^{-8}\,\text{cm}^2/\text{s}$; aggregated polymer present as multilayers $D = 1 \times 10^{-9}$ to $1 \times 10^{-10}\,\text{cm}^2/\text{s}$.

Mixing of LMW emulsifier and polymer (Figure 3.4) increases disharmony and local instability in the plane of the interface, and consequently neither the polymer unification (Woodward *et al.*, 2009) nor the Gibbs–Marangoni mechanisms can work effectively (Wilde and Clark, 1993). This results in a fragmentation of the polymer coverage of the surface and the manifestation (in terms of a thinning of the TLF) of an increase in probe molecule lateral diffusion, as seen in FRAP experiments (Sarker *et al.*, 1995a,b), and a reduction in dilational elasticity. The net effect is an increased likelihood of coalescence of droplets. New approaches use atomic force microscopy to chart the performance of polymers, such as poloxamers and poloxamines adsorbed on the surface of nanobeads (Al-Hanbali *et al.*, 2006; Concannon *et al.*, 2010).

3.3.3 Coalescence processes

Coalescence processes are driven by the rate (frequency) and magnitude of collisions (R), according to Arrhenius. Here the rate is a function of temperature (T) and its effect on droplet mobility, an 'environmental' constant (A) perhaps related to bulk viscosity or network formation, surface fluidity and a collision frequency constant (K) thus:

$$R = A\exp^{(-RT/\ln K)} \tag{3.10}$$

Coalescence is also known as particle fusion and involves the adsorbed layer or membrane rupture. The frequency of collisions and increased kinetic energy imposed by increasing temperature (as in pasteurisation processes) favour fusion. The speed or extent of this process is defined by the sizes of the droplets or particles. Particles with a mass of 300 Da have a bulk diffusion coefficient (D_B) of $130 \times 10^{-11}\,\text{m}^2/\text{s}$ and those with a mass of 59 MDa a D_B of $0.3 \times 10^{-11}\,\text{m}^2/\text{s}$. Particles take an 'Einstein relationship' random walk through the media, but with a diffusional path or distance D_B. The study of coalescence as a function of time is an integral part of product quality control.

Thin liquid (emulsion) films

Considerations concerning a dispersed form of a pharmaceutical product involve the nature of surface coverage of the particle (Figure 3.4) and the physical thickness separating the entities as a result of the TLF that lies between two droplets or particles. The influence of surface coverage is discussed in Sections 3.3.1 and 3.3.2. Thinning of the TLF to an equilibrium value takes place in any emulsion (foam) under the influence of:

- Gravity and its effect on percolation of interstitial entrained liquid.

- Capillary suction of the TLF from the Plateau borders (according to the Laplace equation, which describes internal pressure as a function of size and surface curvature).

- The disjoining pressure in the TLF (see Section 3.2).

It is the integrity and maintenance of integrity of the TLF that keeps droplets away from each other and reduces the likelihood of oil droplet fusion or particle flocculation (Sarker, 2010b).

The equipment used to measure and quantify the TLF thickness (Figure 3.5) relies on internal reflectance of light, interferometry and FRAP (Clark *et al.*, 1991a; Sarker *et al.*, 1995a,b; Sarker, 2010b) of a 300 μm diameter TLF to determine surface mobility/fluidity and curvature (optical interference fringes), equilibrium film thickness (h) and lateral molecular mobility (D; interfacial rheology) within the adsorbed emulsifier layer (Yguerabide *et al.*, 1982; Sarker, 2009b), respectively. The bottom part of Figure 3.5 indicates phenomenologically the internal structure of an aqueous TLF resulting from either polymer or LMW emulsifier surface adsorption. This has been discussed at length elsewhere (Wilde and Clark, 1993; Dickinson and Hong, 1994; Sarker *et al.*, 1995a,b, 1996, 1998a,b, 1999a,b; Sarker, 2005b, 2009b; Woodward *et al.*, 2009) but forms the basis of the establishment of 'thicker' (30–100 nm) or 'thinner' (3–15 nm) TLFs.

Coalescence, the primary mechanism of droplet destabilisation, can be reduced through electrostatic repulsion (by creating a charged surface via adsorption of a charged surfactant) or steric repulsion (by creating a bulky 'rough' surface by adsorbing a large hydrophilic surfactant and an extensive palisade layer). Increasing bulk viscosity merely delays the TLF thinning process, without modification of the final outcome (Sarker *et al.*, 1999). Rupture of the thin film separating droplets or SLN-type particles is common when $\varphi > 0.74$ under the perturbation of thermal instabilities or the fluctuation of mechanical form, which are exacerbated by a fluid or weak interfacial coverage (Castelletto *et al.*, 2003) by LMW emulsifiers. Coalescence means the droplets only get bigger, which is also a result of hydrophobic or interfering 'foreign' particles or aggregates

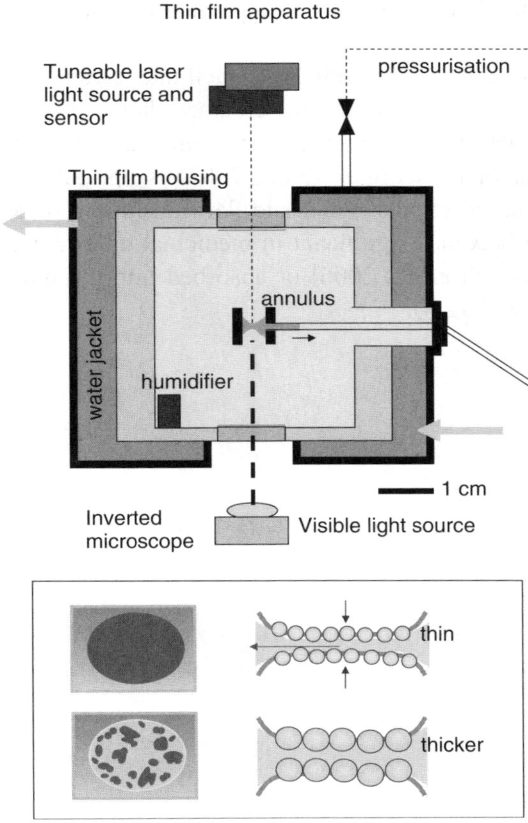

Figure 3.5 Schematic representation of a thin liquid film (TLF) apparatus and the use of surface diffusion experiments (fluorescence recovery after photobleaching, FRAP) and interferometric measurements of an isolated single TLF. The film can be fabricated in either air or oil, mimicking the lamellae of foams or emulsions, respectively

(Sarker *et al.*, 1995a, 1998a, 1999; Sarker and Wilde, 1999; Collins *et al.*, 2008; Genov *et al.*, 2011; Nikolov *et al.*, 2010).

3.3.4 Kinetics of coalescence

Variations in temperature, usually in the form of increases and warming, have a significant role to play in the kinetics of coalescence (see Equations 3.1 and 3.10). Warming of the emulsion and subsequent cooling have two effects. Warming primarily increases the thermal motion of the droplets and, secondly, modifies the interfacial composition and oil density/viscosity within the pharmaceutical emulsion, thus favouring fusion. Phase changes and effective removal of emulsifier following cooling (and crystallisation or aggregation of surface actives) may also be significant.

Lowering of the density in itself may facilitate creaming, which can then lead to large-scale coalescence in the creamed layer. Prevention is possible with higher emulsifier surface coverage (or significant structurisation) and when the polymeric emulsifier (Georgiev *et al.*, 2007) is upright and densely packed (brush form) as opposed to sparse and laid flat (mushroom or pancake form), for example. This also becomes significant in medicinal or parenteral emulsions that are injected (Al-Hanbali *et al.*, 2006) or absorbed into the blood (see Chapter 7 and Sections 5.2.5 and 12.4).

4

Manufacture

The manufacture of an emulsion must be undertaken in a predictable and controlled manner (Sarker *et al.*, 1999), in order to obtain:

- Uniformity of surface coverage.
- Monodispersed droplet sizes in the population produced.
- The smallest droplet size possible.

Manufacture is usually a multistep process, often involving premixing followed by fine emulsification. Hygienic preparation of the product is essential in terms of the environment and the raw materials used (Sarker, 2008). Temperature-induced destabilisation resulting from minute variations is always an issue unless using solid lipid nanoparticles (SLNs) or Pickering emulsions. For this reason, formulators generally use complex mixtures of emulsifiers to form robust interfacial structures (Aulton, 2002) and interemulsifier complexation (e.g. oleyl alcohol and cetomacrogol).

Figure 4.1 shows the general apparatus used for crude droplet size preparation (e.g. ~5–100 µm) or premixing (Figure 4.1a), a French press (high-shear jet) for the production of fine droplets (Figure 4.1b) and a colloid mill for the manufacture of fine droplets (0.1–2.0 µm) (Figure 4.1c).

4.1 Premixing

Blenders of this type (Figure 4.1a) are simple mechanical instruments. Such agitators (e.g. Silverson mixers) usually produce a polydisperse droplet size. Premixing then facilitates micron or nanoscale emulsions in later, more severe emulsification steps (see Table 2.11).

Pharmaceutical Emulsions: A Drug Developer's Toolbag, First Edition. Dipak K. Sarker.
© 2013 John Wiley & Sons, Ltd. Published 2013 by John Wiley & Sons, Ltd.

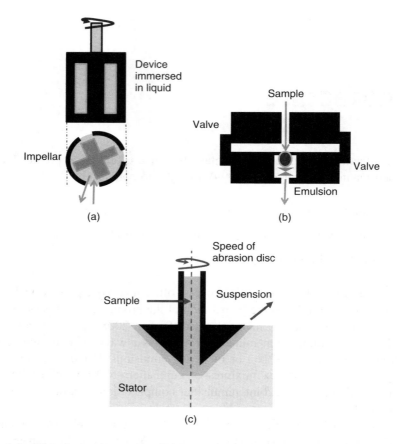

Figure 4.1 High-shear mixers and colloid mills used to impart large shear stresses (MPa) on liquid pharmaceutical samples of varying viscosities (typically 1–3000 mPas) in the presence of emulsifiers prior to settling, filling, packing and storage. (a) Premixer, used to obtain droplet sizes 1–20 μm. (b) French press (high-shear jet), used to obtain liposome/nanoemulsion dispersion or submicron sizes (typically 50–100 nm). (c) Colloid mill, used to give similar submicron (typically 50–100 nm) size dispersions. Combined successive runs can be used to further reduce droplet size in some cases

4.2 High-shear mixers and size reduction

High-shear mixers effect a size reduction by forcing the crude preemulsion through a small orifice under pressure. They are used to perform an initial amalgamation and are usually based on an impellar, turbine or orifice. To be used effectively, they require the inclusion of an emulsifier in the oil/water mix, but this can lead to froth formation due to incorporation of air (e.g. Silverson blenders). The aim is to reduce the median droplet size to around 5–10 μm.

4.2.1 Colloid mill

Colloid mills are used routinely because of their portability and the ability to create pilot-scale batches of product. The ability to impose variable shear in order to obtain fine, moderate and coarse droplet sizes means they are used widely by pharmaceuticists.

4.2.2 High-pressure homogeniser

This equipment (e.g. LabPlant (or PMT) jet mill, French press or avestin emulsificators) may also be used to fabricate liposomes. It uses extremes of pressure (e.g. MPa). Droplet sizes for fine emulsions (nanoemulsions) of 70–200 nm for oil-in-water (O/W) emulsion systems (see Tables 2.2 and 2.11) are possible, depending on the oil fluidity and the shear forces that can be generated. Their value lies in the ability to tune and vary pressure or shear rates in order to determine the end droplet size.

4.3 Multiple and microemulsions

These involve high-shear processes, employing two steps and then combination mixing (see Figure 2.10). This permits the formation of dispersions within a dispersion (Suzuki *et al.*, 1998). For microemulsions, limited shear is usually necessary for fabrication, but a high concentration of multiple emulsifiers is also required (Constantinides, 1995; Pouton, 2000; Bachhav and Patravale, 2009).

4.4 Hot melt (steriles)

One method for ensuring sterility or a pathogen-free product is to use high-quality raw materials, a rapid process following on from mixing and aseptic filling. Hot-melt mixing can be one such route, but it may mean emulsion destabilisation. Filtration (aseptic filtration) subsequent to production under laminar flow conditions and followed by gamma irradiation (^{60}Co) sterilisation is the more customary route to sterile product production.

4.4.1 Pasteurised products

Temperatures of $>70\,^{\circ}$C (pasteurisation) are sufficient to kill vegetative cells (but not the spores from spore-forming microorganisms or extreme thermophiles) confidently and sufficiently to ensure long-term use.

4.5 Filling

Hot filling may be used as part of pasteurisation for externally applied medicines and allows the microbial bioburden (spoilage organisms) to be reduced. Sterile products may be steam retorted at 15 psi or 121 °C, but this always causes instability in the product. Alternative forms of aseptic production are employed (e.g. aseptic filtration following an ethylene oxide-rich production and filling environment and terminal irradiation). However, cold-fill processes can give more stable products physicochemically speaking, although they do not ensure good hygiene. Cold-fill products are used (rarely in pharmacy) where there is no significant risk to injured or damaged skin or of poisoning. Once fabricated, filling of emulsion product usually involves dispensing into aluminium tubes, plastic containers, glass vials, gelatin capsules, transdermal (TD) patches, ampoules and predosed syringes.

II

Forms, uses and applications: biopharmaceutics

The regulatory environment has a larger say in determining the suitability of a product (in terms of the compatibility of the excipients and the active) than most people would suppose (see Table 2.1). Regulatory input is an essential part of modelling product suitability, which is further complicated for complex speciality devices (Gianella *et al.*, 2011; Lawrence and Rees, 2012). For topical products, the Franz diffusion cell—and for solids, the dissolution bath—can go part of the way to mimicking the environment in which the therapeutic is applied. Pragmatically, the constraints of high drug dose, texture limitations, particle size and the restriction to standard medicine forms have a bearing on the use and application of a drug delivery system (DDS). Standard forms of 'emulsion-based' product routinely include:

- Oral liquids (vitaminised products, some gastrointestinal (GI) tract formulations).

- Pulmonary and nasal (insulin, beclametasone).

- Topical (inflammatory: hydocortisone, beclometasone, clobetasol).

- Parenteral (cancer cytotoxics: podophyllotoxin, methotrexate, doxorubicin, paclitaxel; psychiatric: diazepam, midozolam; microbial antifungals/ bacterials: fluconazole, amphotericin, trimethoprim; imaging aids: ^{18}F-FDG, gold nanoparticles (Au-NPs), gadolinium diethylenetriaminepenta acetic acid (Gd-DPTA; gadopentetic acid); anaesthetics: lidocaine, fentanyl, pethidine, propofol).

- Transdermal/matrix (hormones: oestradiol, progesterone; NSAID analgesics: ibuprofen, ketoprofen, indomethacin, diclofenac; smoking addiction: nicotine).

When organised as a mix of formulation types and relevant routes of delivery, the picture is very complex. This part of the book aims to illustrate the ways in which

Pharmaceutical Emulsions: A Drug Developer's Toolbag, First Edition. Dipak K. Sarker.
© 2013 John Wiley & Sons, Ltd. Published 2013 by John Wiley & Sons, Ltd.

products are used. Some products appear under more than one heading and across a multitude of clinical uses. Preformulation development is a crucial preliminary consideration. It must take into account the solubility and biocompatibility in excipients used. Emulsions find use in a range of medicinal products (Sarker, 2008), being administered for internal and external use, and for apolar, toxic (Ruan *et al.*, 2012) and chemolabile drugs.

Preformulation is part of research and development (R&D) and thus a precursor to formulation (Freitas and Müller, 1999; Huailiang *et al.*, 2001; Sarker, 2008). It takes into consideration the physicochemical characterisation of drug pH, pK_a, log *P*, melting point (T_m), glass transition temperature (T_g), crystal habit, polymorphism and so on, and the amalgamation of concepts of bioavailability and bioequivalency with biopharmaceutics. Preformulation involves the application of principles of biopharmacy to the detailed physicochemical parameters of the drug candidate, ensuring they are characterised to achieve an optimum drug delivery. Preformulation of emulsion products must consider the following factors:

- The amount of drug to be solubilised.

- The known chemical properties of the drug.

- The anticipated dose of compound.

- Detailed information for optimal formulation (see Figures 1.1 and 2.2).

In addition to those already indicated, essential characterisation tests would include assay, pH, solubility (range of solvents), buffering, antioxidant requirement, thermal stability (see Figure 3.1), particle size and morphology (see Figure 2.7), rheology, excipient compatibility, isomer racemisation and polymerisation.

5

Creams and ointments

The use of emulsions is very often associated with their therapeutic use as creams and ointments. It may also include emolliency: occlusive films and wetting to facilitate drug passage (see Section 6.1). The range of topical emulsion products that use an emulsified form of drug includes:

- *Cream* An emulsion of oil and water. Penetrates the stratum corneum (SC; outer layer of skin).

- *Ointment* Combines oil and water. Possesses effective skin barrier properties.

- *Gel* Disintegrates on contact with the skin.

- *Paste* Consists of oil, water and powder. Also described as an ointment with a dispersed powder.

- *Liminent (balm, lotion or embrocation)* Not hugely significant for therapeutic use.

A cream is a topical preparation ($\sim$15–50% oil, concentrated oil-in-water (O/W) emulsion), usually meant for application to the skin (Sarker, 2006b)—the 'largest' organ in the body (SC, ear)—and to key mucus membranes (nose, vagina, rectum). Creams are used across a broad range of skin conditions (eczema, psoriasis, acne vulgaris). They are referred to as 'semisolids' and often use a thickener or stabiliser. Various types of emulsion are described in Table 2.2. O/W types are by far the most common (Huailiang *et al.*, 2001) and well-tolerated. Water-in-oil (W/O) emulsion types are also used often, generally for emolliency purposes. They produce occlusive films, reducing water loss from the skin. Creams have useful barrier properties which can protect the skin, and can also be used as vehicles for drug substances (antibiotics, antifungal drugs, local anaesthetics, antiinflammatory drugs, antipruritic drugs) when dispersed in a suitable base.

Commonly used examples of creams based on O/W emulsions (see Table 5.1) or aqueous microcrystalline dispersions of long-chain fatty acids include those administered via the vaginal route (e.g. sulphacetamide + sulphathiazole + sulphabenzamide cream).

Pharmaceutical Emulsions: A Drug Developer's Toolbag, First Edition. Dipak K. Sarker.
© 2013 John Wiley & Sons, Ltd. Published 2013 by John Wiley & Sons, Ltd.

Table 5.1 Creams, ointments and transdermal products[a]

Type	Name	Drug	Use
Cream	Orudis	Ketoprofen	Analgesia
Cream	Advil	Ibuprofen	Analgesia
Cream	Lanacane	Benzocaine	Anaesthesia
Cream	Lamisil	Terbinafine	Anti-fungal
Cream	Dermovate	Clobetasol propionate	Dermatitis/eczema
Cream	Anusol	Zinc oxide + bismuth oxide (astringents) balsam Peru (antiseptic)	Haemorrhoids
Ointment	Accutane	Isotretinoin	Acne
Ointment	Betnovate	Betamethasone valerate	Dermatitis/eczema
Ointment	PremarinS	Oestradiol	Gynaecological
Transdermal patch	Qutenza	Capsaicin	Analgesia
Transdermal patch	Nicorette	Nicotine	Replacement therapy

[a]Thermodynamically unstable and susceptible to phase separation.

Basic ingredients for eczema O/W cream ($\sim$30% paraffin; 2–4% emulsifier) are:

- water;

- 10% liquid paraffin;

- 15% paraffin wax;

- 5% beeswax and microcrystalline wax;

- butylated hydroxyanisole (BHA);

- sorbitan sesquioleate.

Basic ingredients for fungicidal O/W cream (22% wax and long-chain alcohols; 2–4% emulsifier) are:

- benzyl alcohol;

- polysorbate 60;

- sorbitan stearate;

- 15% cetyl esters wax;

- 5% cetyl stearyl alcohol;

- 2% 2-octyldodecanol;

- water.

Creams may represent the most popular and traditional use of emulsions. They are favoured because of their inherent spreadability.

An ointment is often an oil-based (25–50% for O/W emulsions and >80% for W/O adsorption bases) dosage form, intended for use topically on a variety of surfaces, such as the skin (including hands and feet) and the mucous membranes of the vagina, penis, anus/rectum, lips, nose and eyes. The vehicle (dispersion medium) of an ointment is referred to as its 'base'. The selection of a base depends upon the end use of the ointment. Different types of ointment media include:

- hard and soft paraffins, various grades;
- emulsifying wax blends and cetrimide;
- beeswax, plant wax, wool fat;
- macrogols (poly(ethylene glycol), PEG);
- vegetable oils.

The drug is distributed in the base as a proper thermodynamic mixture or dispersion. Ointments are generally used as emollients or for the application of actives to the skin (Table 5.1), as well as for protective purposes where occlusion is required and miscibility with skin secretions is anticipated (Florence and Attwood, 1998; Benson and Watkinson, 2012). Parameters controlling drug uptake from the vehicle (Lipinski parameters) to the blood plasma (via the skin) are discussed in Section 12.2. Ointments are fabricated by 'trituration', where micronised active pharmaceutical ingredients (APIs) are dispersed by grinding with increasing amounts of base (as a diluent), or by melting together the components in the order of their melting points, ensuring effective mixing and an even consistency.

5.1 Nutraceuticals and cosmeceuticals

Nutraceutical foods, supplements and diet products are based on emulsions of vegetable oil, which may include α-tocopherol, hydroxypropyl (methyl cellulose) (HPMC), magnesium stearate, Span 80, corn oil (vehicle), retinoyl acetate and ascorbyl palmitate as a corn-oil O/W emulsion system (D'Ascenzo *et al.*, 2011). Depending on the dosage form chosen (tablet, chewable pastille, liquid syrup or food product), the emulsion and water-soluble vitamins are dispersed in a complex mixture and the broader product.

5.2 Medicinals

5.2.1 Dermatoses, eczema and corticosteroidals

The drugs and therapeutics typically used in formulations include retinoids, isotretinoins, cyclosporin, steroids and fat-soluble vitamins. Topical steroids have different classes defined by their therapeutic potency:

- *Lower potency* Cortisone acetate, tixocortol pivalate, hydrocortisone, prednisolone, methylprednisolone and prednisone.

- *Low–medium potency* Mometasone, budesonide, triamcinolone acetonide, amcinonide, desonide, fluocinonide and halcinonide.

- *Medium–high potency* Dexamethasone, betamethasone and fluocortolone.

- *High potency* Prednicarbate, aclometasone dipropionate, clobetasone-17-butyrate, betamethasone valerate, fluocortolone caproate and fluprednidene acetate.

The potency of a therapeutic agent defines its use in terms of the location and duration of treatment.

5.2.2 Acyclovir and antivirals

Antiviral products are often O/W dispersions with the apolar active (e.g. acyclovir) encapsulated in an oil droplet. Products are used topically and parenterally (but also nasally and via the pulmonary route) across a range of therapies. Some typical products include:

- Ritonivir (Norvir, anti-HIV) emulsion paste in soft gel capsules.

- Acyclovir (Zovirax, Herpes zoster).

- A whole range of vaccines (see Table 1.2) that have been developed or are in development for application locally to this region.

Vaccines (see Figure 2.2) have been made using an immune-influential peptide from a lymphocytic choriomeningitis virus (LCMV), along with block copolymer in the form of an emulsion vaccine. Many others are currently in development (Huailiang *et al.*, 2001; Bivas-Benita *et al.*, 2004; Rudolph *et al.*, 2004; Csaba *et al.*, 2009).

5.2.3 Genitourinary and rectal uses

These products make use of pessary implants for male and female penile/vaginal/urethral therapy or of suppositories for rectal delivery. A dispersion of gas-in-emulsion as a foam may be employed, in a similar way to a simple O/W emulsion. Complex foams (air-in-(oil-in-water)) of this type include Betamousse, a colonic inflammation product (betametasone) used for irritable bowel syndrome (IBS) and related ailments. Penile and vaginal antibiotic creams, lotions and ointments also exist, as do alprostadil urethral suppositories, used for the treatment of male erectile dysfunction. A selection of creams (lidocaine + zinc oxide + fluconazole) are used to counteract yeast infection (thrush). Rectal hydrocortisone products are formulated for the treatment of haemorrhoids in the form of a cream and suppositories, along with progesterone and oestrogen creams for hormone deficiencies and local effects (Jones and Harmanli, 2010).

5.2.4 Antifoams

Simeticone (dimethicone; silicone oil), a low-hydrophile–lipophile balance (HLB) surfactant/oil, is used in oil-laden dispersions such as Infacol, in which HPMC (hypromellose polymer for a Pickering-aided dispersion of silicone oil) is used as the polymeric emulsifier and surface stabiliser (Concannon *et al.*, 2010). Dimethicone is used to treat neonatal acid reflux resulting from milk protein froth formation in the stomach (or indeed during peptide synthesis biotechnology processes). These oils function by forming hydrophobic solids, which destabilise aggregates (Sarker *et al.*, 1996) or oil droplets (lenses) that sit in an aqueous thin liquid film (TLF) and prevent continuity and the polymer or Gibbs–Marangoni stabilisation mechanisms (see Figure 5.1). Occasionally, these oils are mixed with trisiloxane spreading agents, which are used industrially to aid surface wetting (Rafai *et al.*, 2002).

5.2.5 Miscellaneous

All shampoos are mostly water, with 2–8% detergents (and fatty alcohols) and foaming agents and about 1% fragrance and preservatives. Oils are often present

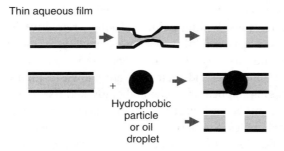

Figure 5.1 Thin-film rupture mechanisms involving wetting, adhesion, lensing and lamellar disjoining. This behaviour is typically used to break foams (e.g. neonatal gastric milk froth). It is also used in a biotechnology context, with antifoam surfactants and oils or emulsions composed of these substances

in O/W microemulsion or O/W emulsion form (Sarker, 2008). Shampoos often contain antitangling (cyclomethicone/simeticone oil) agents, as well as viscosifiers (thickeners, e.g. cellulose), humectants (glycerol), sequestering agents (EDTA), colour (microemulsion-encapsulated dyes) and conditioners (glycol distearate; coconut-oil-derived emulsions and wetting agents). Routinely used detergents include sodium lauryl sulphate, laureth sulphate and sulphosuccinate (AOT) in micellar/microemulsion form. Medicated shampoos (e.g. Nizoral with the antifungal ketoconazole) may contain denaturing salicylic acid to loosen flakes of skin or dimethicone as a wetting/smoothing/occlusion aid (the silicone film intercalates within hair cuticles and reduces friction and hair breakage, maintaining colour and gloss).

Dimethicone (also used in treatment of head lice) and selenium sulphide, zinc pyrithione, ketoconazole and ciclopirox are all employed on the scalp to reduce the number of yeast cells. Broad-spectrum biocides include isothiazolinone (Kathon CG-types) and the related methyl-isothiazolinone and methyl-chloroisothiazolinone. When the protective scalp-surface oil film (sebum) is removed, surfactant degreased skin may become excessively dry and sore and produce contact dermatitis.

Fluorinated emulsions

Perfluorocarbons (PFCs) are derived from hydrocarbons, but with the hydrogen substituted by fluorine. Examples include octafluoropropane, perfluorohexane and perfluorodecalin but may also extend to fluorinated block copolymer emulsifiers. PFCs are favoured in products for their thermal stability, colourless appearance, greater density than water, lower viscosity than water, low surface tension, volatility and chemical inertness (Courier *et al.*, 2004). They mix well with hydrocarbons such as hexane but are not miscible with common solvents and water. They are exemplary solvents for gases, such as oxygen, because of their very low intermolecular forces. The physical properties of PFCs are defined by the molecular mass of the carbon backbone. Larger carbon chains unsurprisingly give higher viscosities, surface tensions, boiling points, densities, refractive indexes and vapour pressures. However, gas solubility decreases notably as the number of carbon atoms increases, and this may limit their applicability. Fluosol, Oxygent and Lipiodol halogented emulsions and their uses are shown in Figure 5.2. Therapeutic uses include imaging applications, chemotherapy and tissue repair/replacement.

PFCs are used in *in vivo* ultrasound reflection, forming density-contrasting aphrons and nanobubbles. Similarly, they have a role in magnetic resonance imaging (MRI) using fluorine-19 nuclei (^{19}F) and x-ray radiography (e.g. perfluorooctyl bromide (PFOB)) as a contrast agent, due to their raised density to photons and thus opacity. Fluorine is not present at a significant concentration in the body and can thus be located easily in positron emission tomography (PET). PFCs can dissolve around 30% v/v oxygen but are too dense for routine pulmonary use and

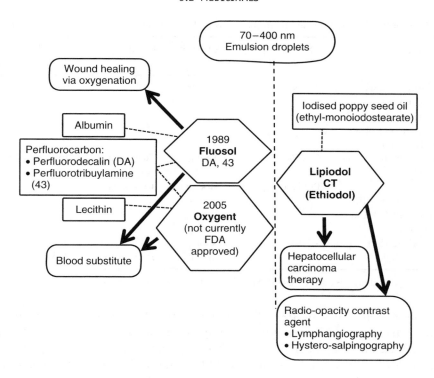

Figure 5.2 Fluorinated and other emulsions in commercial and varied pharmaceutical and medical use

therefore find a role as blood substitutes in the form of 'artificial red blood cells' (Riess and Krafft, 1998; Courier *et al.*, 2004; Krafft and Riess, 2009).

Fluosol-DC (or Alliance Pharmaceutical Corporation's lecithin-stabilised product Oxygent), for example, has the approximate following ingredient composition:

- perfluorodecalin, 25.0% w/v;

- egg yolk phospholids (or F68 block copolymer emulsifier), 3.6% w/v;

- fatty acid(emulsifier), trace;

- d-sorbitol (stabilizer), 3.5% w/v;

- NaCl, 0.20% w/v;

- KCl, 0.01% w/v;

- $MgCl_2$, 0.01% w/v;

- sodium lactate, 0.11% w/v.

Table 5.2 Thermal transitions in the lipids and lipid fractions of common emulsifiers

Emulsifier type	Melting point, T_m (°C)	Solubility	Polymorphism	Enthalpy of melting (kJ/Mol)
Lecithin (various mixtures of myristate, palmitate, stearate and oleate fatty acids)	25–60	Apolar	–	60–75
Glyceryl monostearate (GMS)	58–59	Apolar	Two forms	150
PEG400 monooleate – 10 (a.k.a. PEG-8 ester)	PEG400 $T_m = 3$	Moderate polarity	Two forms	45
Span (80)	−20	Apolar	–	325
Polysorbate (80)	−20	Apolar	–	300
Oleyl alcohol	13–19	Apolar	Two forms	273

PEG, poly(ethylene glycol). PEG400 has a molecular mass of ~400 Da.

With a droplet size of 100–200 nm, a fraction of the size of an erythrocyte (5 µm), these droplets can permeate erythrocyte clusters in arterial circulation and defy clearance by the reticuloendothelial system (RES; monocyte–phagocyte system, MPS). These products can permit blood transfusions free from bloodborne diseases for all blood groups.

5.2.6 Key excipients and ingredients

Preformulation of key excipients and formulation aids should create a product without significant inconsistency and with poorly predicted behaviour (Table 5.2). For all emulsion products, the oil fraction is usually composed of paraffin, silicone or purified food oils (triglyceride, olive, safflower, squalene, coconut, palm, peanut, castor and soya oil) and waxes. Emulsifiers and solubilisation aids include propylene glycol (penetration enhancers), polymeric emulsifiers, fats, wax, alcohols and synthetic surfactants. Other ingredients (Busse, 1978) include bulk phase stabilisers/viscosifiers, preservatives and antioxidants and purified water. Essential characteristics of all formulation aids and primary ingredients are excipient biocompatability and minimal toxicology.

6

Pastes and bases

Water-in-oil (W/O) bases (hydrophilic and nonhydrophilic) usually have a low surfactant content (see Table 5.2) but a significantly high oil content and are generally used to create occlusive films. Pastes by definition have a high solids content. The solid fraction dispersed in oil (solid-in-oil base) or an O/W emulsion oil droplet dispersed in water with stabilisation via inclusion of solid, such as bentonite magma, calcium phosphate ($Ca_3(PO_4)_2$), silica (SiO_2) or starch, provide a means to create a stabilising texture. A paste behaves like a plastic material (Florence and Attwood, 1998; Jones, 2002; Sarker, 2010; Desi Reddy *et al.*, 2012), with an interlocking disordered structure, until a sufficiently large load (or heat, as in the case of a suppository) is applied (Acartürk, 2009), whereupon it flows like a fluid.

6.1 Emolliency

Wetting and spreading of ointments and paraffin wax products on the skin in order to provide an occlusive film is used against the drying and chafing found with eczema, psoriasis and other dry skin disorders manifested in irritated and broken skin. Examples of emollient products include:

- Cetraben cream, ~23% O/W dispersion (Genus Pharmaceuticals).

- Epaderm ointment, >80% W/O dispersion using heavy paraffin (Vaseline) as base (Molnycke Healthcare).

- Diprobase cream, ~23% O/W dispersion (Schering-Plough); Diprobase ointment, >80% W/O dispersion using heavy paraffin as base (Schering-Plough).

Ointments have the advantage of persisting longer on the skin and being less easily rinsed off than absorption bases or creams. Emollient products of this type are all liquid/heavy paraffin (various viscosities) emulsions. These products also routinely contain propylene glycol (wetting agent), simeticone (occlusive film former

Pharmaceutical Emulsions: A Drug Developer's Toolbag, First Edition. Dipak K. Sarker.
© 2013 John Wiley & Sons, Ltd. Published 2013 by John Wiley & Sons, Ltd.

and spreading agent), olive or soya (soybean) oil, emulsifying wax, cetostearyl alcohol, macrogol, glycerol and preservatives (e.g. parabens, E216). Bath oils are simple formulas which contain liquid paraffin, isopropyl palmitate and an amalgam (Acartürk, 2009), again used to maintain skin hydration (see Chapter 8).

6.2 Suppositories

Suppositories are used for direct therapeutic purposes (e.g. paracetamol, diclofenac and opiate suppositories) or for laxative purposes (enema). Under general medical administration the active substance crosses the rectal mucosa (Acartürk, 2009) into the bloodstream following disintegration/melting of the vehicle/base and spreading of the drug particle along the wall of the void (Chaiseri and Dimick, 1986). This is discussed in Section 2.1.2, which covers theobroma oil chemistry and the characterisation of simple lipids and cocoa butters. To avoid the effects of fat polymorphism (see Table 2.5) and variations in crystal form, suppository fats such as cocoa butter are first heated to 50–52 °C, which eliminates their crystal 'memory', then cooled to 25 °C, which establishes the type I–III forms, and finally tempered at 33–35 °C, which allows type IV to predominate. Using Span 60 or Span 65, Tween's or propylene glycol surfactants at 1–10% in a sample maintains the type IV and V forms as a liquid fraction and eradicates excessive coarse crystal formation. Some suppositories are made of poly(ethylene glycol) (PEG) and glycerol rather than solid fat (e.g. a PEG base 400/4000 mixed with glycerol (~70%) and with gelatin (~15%) for a water-soluble base). In either case, the vehicle is a heat-sensitive composite material of a fatty or lipid-like dispersion (Table 6.1). Other actives routinely encapsulated in suppositories include terconazole (antifungal), oestradiol/progesterone (hormones), valium (anxiolytic), clindamycin (antimicrobial) and corticosteroidals (antiinflammatory).

Table 6.1 Phase changes driven by monotropic polymorphism and its influence on melting and softening transitions

Carbon	Lipid	α	β'	β
C12 : 0	Trilaurin	15 °C	34 °C	47 °C
C14 : 0	Trimyristin	33 °C	45 °C	59 °C
C16 : 0	Tripalmitin	45 °C	57 °C	66 °C
C18 : 0	Tristearin	55 °C	63 °C	74 °C
C18 : 1	Triolein	−32 °C	−12 °C	6 °C

β-forms of fats have a lower number of defects (therefore expulsion of drug is better), show the highest T_m, have the most stable (chair) conformation and provide the largest crystals (100 μm) compared to the 5 μm α-forms.

6.3 Pessaries

These drug-bearing products are suppository-like (rather than surgical devices) porous (sponge or solid foam matrix) or composite globular, egg-shaped or conical media ($\sim$5 cm^3) for routine vaginal insertion. Mixtures of hard fat and oil and nanoemulsions (Henzl, 2005; Jones and Harmanli, 2010) are used routinely as the base, for example Witepsol or Suppocire fats (Acartürk, 2009). Frequently, these body-dispersible types (e.g. the Prostin E$_2$® dinoprostone (prostaglandin E$_2$) vaginal suppository (Pfizer)) use an amalgam of paraffins (liquid/soft), glycerol, gelatin, water, lecithin and triglycerides as oxytocic drug delivery systems (DDSs) (and thus occasionally an abortifacient). When not fat-based, they are often plastic- or silicone-impregnated devices (see Section 8.5), which are inserted into the vaginal canal (or rectum). Examples include:

- Cervagem–Lamicel (abortive);

- oestrogen/oestradiol for uterine prolapse;

- sintocinon (oxytocin) birth-inducing hormone;

- Cervidil prostaglandin E$_2$ for labour inducement;

- antifungals (terconazole, fluconazole, clindamycin);

- antiinflammatories (steroidals).

Optimal drug delivery occurs when formulators choose a base in which the drug is *least* soluble, thus facilitating optimum drug release to the vaginal membrane. The ideal vehicle should therefore:

- melt at body temperature;

- dissolve in body fluids;

- be significantly nontoxic and nonirritating;

- be compatible with any drug also applied at this location;

- release the active pharmaceutical ingredient (active) readily.

An *in vitro* survey profiling the potential vaginal mucosal permeation of a proposed anti-HIV microbiocide and the influence of intrinsic formulation excipients on its efficacy has been reported recently; it serves to illustrate the possibilities for more widescale use of this delivery route (Grammen *et al.*, 2012).

7

IV colloids

Many drugs are best suited to intravenous (IV) administration. This allows rapid, immediate and high, yet predictable, metered dosing. The vast majority of these IV drug products are colloidal in nature, including emulsions, polymer micelles, microemulsions, nanoemulsions and liposomes (Sarker, 2005a, 2006a,b, 2009a, 2012a; Al-Hanbali *et al.*, 2006; Concannon *et al.*, 2010). A capsular form is favoured as it permits an increased drug load, better release profile, reduced toxicity and increased chemical stability (Almeida and Souto, 2007). These parameters are described in Table 7.1. In most cases, both increased drug solubility and enhanced circulatory persistence favour the nanoscopic emulsion and colloid drug delivery system (DDS).

Liposomal products (Table 7.1) are both diverse and highly standardised in terms of structure and application. Commercial vesicular products of this type include Doxil, a 'stealth' (macrophage-avoiding) mPEG-DSPE (PEGylated lipid) liposome (and dual-content Caelyx), AmBisome, DaunoXome and the earliest Food and Drug Administration (FDA)-approved product, Taxol (1992). Other products (Maurer *et al.*, 2001), such as Myocet (Sopherion Pharma), are only formally approved for use in the EU/Canada (2004). A fuller list of liposome-encapsulated therapeutics is given in Table 7.2. The SPI-77 PEGylated stealth dosage form, an mPEG-DSPE liposome with cisplatin, used for ovarian cancer (CA), is currently in phase II clinical trials. Liposomes embedded with superparamagnetic iron oxide (SPIO) nanoparticles (particles cause drug release (leakage) on heating in a radio-frequency electromagnetic field) are in development. In the majority of cases, drug is 'emulsified' and encapsulated in an aqueous core. Stealthification and the possibilities for bilayer sequestration of lipophilic drugs have been discussed (Sarker, 2009a,b); this may give another form of emulsification to high-log P moieties.

Niosomes are vesicular systems made from nonionic emulsifiers (Sarker, 2012a; Kazi *et al.*, 2010), such as Span 80. Although few approved drug products currently exist, this is likely to change significantly over the next decade given the potential of these DDSs. Current products include:

Pharmaceutical Emulsions: A Drug Developer's Toolbag, First Edition. Dipak K. Sarker.
© 2013 John Wiley & Sons, Ltd. Published 2013 by John Wiley & Sons, Ltd.

Table 7.1 Application and therapeutic benefits of using nanoparticle drug delivery systems (DDSs)

Form	Type	Size	Example(s)	Use
Liposome, 1970s	Stealth, SUVs IUVs	~25 nm	PEG–liposome + doxorubicin (100 nm)	Enhanced circulatory persistence
Micelles	Polymer and emulsifier	20–100 nm	Poloxamer/ poloxamine (Pluronic/ Tetronic)	Improved drug solubility
Emulsions and microemulsions (μem)	Nanotemplate Engineering Technology (μem)	Submicron, SLN ~50 nm	Aqueous (O/W), (S/W) drug	Improved drug solubility

SUV, small unilamellar vesicle; IUV, intermediate unilamellar vesicle; SLN, solid lipid nanoparticle (a type of emulsion); O/W, oil-in-water; S/W, solids-in-water.

Table 7.2 Liposomal drug delivery and the range of drug molecules encapsulated

Encapsulated drug	Disease	Liposomal delivery benefit
Amphotericin B	Fungal infection	Reduced kidney damage
Doxorubicin	Cancer	Reduced cardiotoxicity, immune damage, emesis, alopecia
Cisplatin	Cancer	Reduced emesis and kidney damage
Vincristine	Cancer	Reduced kidney damage
Gentamycin	Gram-negative bacterial infections	Reduced kidney damage
Indomethacin	Arthritis	Reduced gastric toxicity
Philocarpine	Glaucoma	Reduced dosing
Indium (111)	Tumour imaging	Preferential accumulation in tumour
Epithelial growth factor	Wound healing	Infrequent dosing

- Visudyne (verteporfin), for ocular photodynamic therapy (PDT) in connection with macular degeneration (FDA approved, 2000).

- Sporonox (itraconazole), an antifungal (FDA approved).

- Diallyl sulphide (food/drug additive, FDA approved).

- Methotrexate (antineoplastic), in phase II clinical trials.

- Resovist (ferucarbotran), used for magnetic resonance imaging (MRI), in phase II clinical trials.

- Gadomer 17 (gadolinium complex), used for MRI (radiotherapy), in phase III clinical trials.

Micellar (normal rather than reverse/inverse micelle) systems are based on self-assembled colloidal particles. The core of the particle is usually hydrophobic (see Tables 2.8 and 2.9), due to assembly driven by entropy and the 'hydrophobic effect'. In many senses, the core of a micelle can be considered to be 'oil-like' and therefore represents a form of emulsion. Most therapeutic micelles are formed from polymeric emulsifiers (see Tables 2.8 and 2.9). This is even more likely when the product has been/must be rigorously treated to ensure sterility or is exposed to variable pH. Some recent examples of micellar products (see Tables 1.2, 2.9 and 7.1) include:

- Genoxol PM (paclitaxel) poly(ethylene glycol) (PEG)-phospholipid polymer micelle (Samyang Co, Korea), currently in phase II clinical trials.

- PluroGel, a Pluronic P105 crosslinked micelle with silver sulphathiadine (PluroGen Therapeutics) for wounds.

- PEO-PPO aqueous polymeric micelle, for housing 1.5 wt% Efavirenz vaccine (HIV/AIDS), currently in phase I clinical trials.

- Block copolymer micelle (plasmid DNA) base, for the DermaVir matrix trans-dermal patch for HIV/AIDS (Dream Ventures Inc.), currently in clinical trials.

- NBI-29 radioactive (lyophobic gold-based) micelle, for prostate cancer (Shashun/NBI, India), currently in trial.

- Surfaxin (lucinactant pulmonary peptide micelle) aerosol, for respiratory distress syndrome (RDS) in neonates (Discovery Labs), awaiting FDA approval.

- CYT-6091 Aurimune radioactive colloidal PEG-gold (CytImmune Sciences Inc.), currently in phase II clinical trials.

- Jevtana (carbazitaxel) in combination with steroids, approved by the European Medicines Agency (EMEA) and the FDA (Sanofi-Aventis) for advanced prostate cancer.

- DACH-Platin-MediCelle, for cancer; this is Eloxatin (oxaliplatin, from NanoCarrier) housed in a PEG-polyglutamic acid block copolymer micelle using Debiopharm's platform micelle technology, currently in preclinical trials.

The nanoemulsion represents a huge growth area and adaptation of the coarse emulsion. A combination of multiple emulsifiers (Span, Tween, poloxamers, lipids) and an improvment in manufacturing equipment means 40–100 nm emulsion droplet manufacture is now routinely possible. Efficacy in this case is vastly improved, along with surface area and penetration power (Gregoriadis, 1973, 1977; Torchilin, 2001; Sarker, 2012a). Currently approved nanoemulsion therapeutics and products (Almeida and Souto, 2007) in development include:

- NanoStat NB001, an antiviral product (NanoBio), currently in phase III clinical trials.

- Rective, a nitroglycerine ointment (Prostraken) for anal discomfort (haemorrhoids/piles/tears), FDA approved (2011).

- Intralipid, a nutrition (Pharmacia/Kabi) lipid depot, FDA approved in 1972, EU approved in 1962.

- Ropion, a flurbiprofen axetil prodrug nanoemulsion (Kaken) and analgesic, FDA approved.

- Palux, a lipid nanosphere platform (for the perfumery industry), FDA approved.

- Liple with alprostadil, a vasodilator for erectile dysfunction, FDA approved.

- Limathason containing dexamethasone palmitate (Mitsubishi Pharma), FDA approved.

- Diprivan with propofol, used for anaesthesia (AstraZeneca), FDA approved.

- Omnitrope, a human growth hormone suspension for injection (Sandoz), FDA approved.

Solid lipid nanoparticles (SLNs) are now widely exploited for cancer therapeutics (e.g. camptothecin (as far back as 1991) and 5-fluorouracil), but also for hormones such as testosterone, analgesics such as ketorolac (a nonsteroidal antiinflammatory drug, NSAID) and antifungal drugs such as itraconazole.

Parenteral products (e.g. liposomes and nanoemulsions) mainly target the systemic routes (by means other than through the digestive tract, but especially by injection, e.g. eye, blood, neural tissues, lymph, subcutaneous fat, intrasynovial fluid, spinocerebral fluid, etc.). The targeting is purposeful. To most scientists, this usually implies a dosage form made by invasive injection, but these products are by no means restricted to this. IV or intradermal (ID; depot) and intramuscular (IM) or intrathecal (IT; lumbar puncture) are the most common methods of application. Parenteral emulsions of Lipiodol with specific active pharmaceutical ingredients (APIs) may include pirarubicin (gastric cancer), epirubicin (liver cancer), cisplatin (prostate cancer), pingyangmycin (intestinal cancer), human recombinant tumour necrosis factor (liver cancer) and methotrexate (cervical cancer). Chemotherapy is not restricted to the antineoplastic drugs but can also include antimicrobial drugs (e.g. amphotericin, lopinavir, ritonavir, azithromycin, penicillin) and corticosteroids (e.g. halcionide, clobetasone, hydrocortisone-17-valerate). Essential criteria for such products are a sub-half-micron average particle size with a very narrow size distribution and freedom from phase inversion. Emboli formation in narrow vessels can represent a possible problem where droplet size is bigger than an erythrocyte.

Long circulating emulsions are made by covering sub-100 nm oil droplets with a 'dense and voluminous' hydrophilic polymer or emulsifier (bilayer) 'skin' (see Figure 7.1) of differing physical characteristics. This permits longer blood residence and avoidance of capture by monocyte–phagocyte systems (MPSs) (Al-Hanbali et al., 2006) macrophages in the spleen, liver, lung, blood, bone marrow and so on (Sarker, 2012a). Increasing circulation is made possible by extensive use of large (<2 kDa) hydrophilic polymers or at least polymers in which the palisade layer is hydrophilic, and often by replacing natural lecithin emulsifiers with poloxamers (e.g. 388 or PEGylated amphiphiles) and nonionic emulsifiers (e.g. Tween 80). Post-administration, fine emulsions (and liposomes) in the circulatory system acquire apoproteins from circulatory high-density lipoproteins (HDLs). Breakdown and thus drug release are limited by endogenous factors such

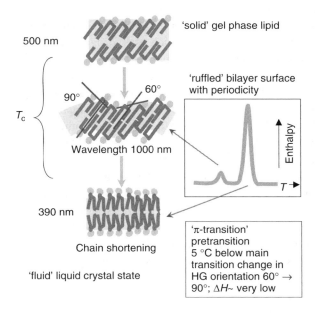

Figure 7.1 Thermal transitions (by differential scanning calorimetry, DSC) in the liposome/vesicle or micelle/lipid particle core. The figure illustrates changes which polymer–lipid or lipid molecules such as low-hydrophile–lipophile balance (HLB) surfactants/emulsifiers and phospholipids typically undergo. Phospholipids typically show a premelting π-transition

as compositional variation, but with very high dosages the excess lipid can impair reticuloendothelial system (RES) function with suppression of the immune system. This is usually not the case for medical dosages of therapeutics, where moiety potency negates a significant volume of lipid, but it can occur for IV nutrition. Some parenteral nutrition bases have been used as the basis for dosing anaesthetics and other drugs (e.g. Medialipide (B.Braun) and Intralipid (Kabi) for ropivacaine, bupivacaine and levobupivacaine). General anaesthesia by IV injection means administration of actives such as methoxyflurane (dose $\approx$ 100 ml of 5–10%), 'intralipid emulsions', halothane and other volatile anaesthetics (Sarker, 2008). The advantage of parenteral emulsions is a typically more uniform droplet size, surface charge and surface composition, and because of this a far superior efficacy and dosage management. This permits a degree of flexibility in terms of injection site administration and the admixtures metered directly into a timed syringe system or polyfusor bag for use via an IV line (PICC, Hickman or Groshong central line).

7.1 Needle free

Companies such as Powderject, Bioject and Aradigm produce a range of implants and implantation technologies that circumvent use of a hypodermic syringe. These types of product are gathering momentum, related to the large-scale inoculation/vaccination programmes favoured by the World Health Organization (WHO) and others and to use of adaptations such as hydrogel–emulsion composite (e.g. Exubera, insulin via the pulmonary or nasal routes). Other novel delivery devices arise from far-reaching contemporary industrial collaboration (e.g. Nektar Therapeutics and Pfizer) and are based on dispersion/emulsion-bearing insulin and other hormones. Products such as those from Bioject are capable of delivering a large number of existing drugs without the need for reformulation (e.g. Furzeon for HIV treatment, from Roche). Still other products that have been recently clinically tested (clinical trials, CTs) include Medi-jector vision (Antares), Intraject (Aradigm), Crossject (CMT), Miniject (Biovalve) and PenJet (PenJet Corporation).

Over-the-counter (OTC) anti-allergy nasal sprays contain steroids such as Beconase (beclometasone) and Flonase (fluticasone) formulated in thickened, micellar (Tween 80) 'solution' form. Anti-flu vaccines such as FluMist (Aviron) are again based on a polymeric emulsifier system for nasal application. There are pragmatic reasons for and real advantages of microemulsion and micellar nasal spray use:

- A large range of molecules.

- Noninvasive application.

- Rapid uptake of the drug itself.

- Metabolic degradation avoidance.

- Therapeutic access to the brain (taking into account the blood–brain barrier, BBB), which can be practically impenetrable to coarse and even finer emulsion products.

Examples of products include diltiazem, used for hypertension and appetite suppression (Compellis Pharmaceuticals), and OptiNose, a benzodiazepine (somnolent) from MedPharm.

Microneedles represent a new and interesting use of multiple cellulose- and/or polymer-type composites for the introduction of vaccines and may also be seen as nanopatches for influenza (Alza Corporation). Microneedles (needles average from gauge 21 to 27, or 0.8–0.4 mm diameter) and patches based on microneedle (100 µm-diameter bore) technology (cellulose) dissolve *in situ*. Larger-bore needles are felt by the patient upon application. Examples of microneedle producers of wearable patches containing vaccines include Biovalve, Sanofi-Pasteur, Becton-Dickinson and Co. and 3M (Huailiang *et al.*, 2001). The lymphatic system offers preferential take-up of emulsions from microneedle systems, and this favours treatment of certain types of cancer, including lymphoma, melanoma and leukaemia.

The pulmonary route, via a metered-dose inhaler (MDI), can be used for a range of solids and SLNs (Videira *et al.*, 2002; Bivas-Benita *et al.*, 2004, Mansour *et al.*, 2009). Examples of products that promote micro- and nanosphere technology platforms for pharmacy use include Promaxx (Epic Therapeutics), which includes encapsulated actives such as antitrypsin (a treatment for chronic obstructive pulmonary disease (COPD) and emphysema), developmental hormones, short interfering RNA and antisense oligonucleotides (a treatment for cystic fibrosis, via gene therapy) in the form of microspheres, which are constituted from PEG, polysorbates and lipid polymers. These surface-active agents act as solubilisers to aid in particle encapsulation for direct delivery to the lungs.

7.2 Ocular therapy

Ocular therapies (and otic) often use SLNs, micelles or microemulsion formulations (Cavalli *et* al., 2002) where thermodynamic stability, nanoscale particle kinetics and transparency are desirable for surgical anaesthesia (e.g. lidocaine, bupivacaine). Other actives routinely delivered to the eye include gentamicin, ketoconazole, miconazole, neomycin, ofloxacin, polymyxin B and tetracycline, all for antimicrobial purposes (Souto and Muller, 2005; Bhalekar *et al.*, 2009). Corticosteroidal drugs such as dexamethasone, betamethasone, fluocinolone, hydrocortisone or prednisolone and antihistamines are used regularly in an antiinflammatory capacity. Micellar and microemulsion eye drops are formulated in aqueous saline in order to administer pharmaceutical actives to the eye.

7.3 Cancer

Cancers require rapid blood vessel development (Primo *et al.*, 2007; Davis *et al.*, 2008; Young and Michelson, 2012) and a rich supply of blood to fuel their growth. Cancerous tissue is 'leaky' as a result of this rapid 'disorganised' growth. This can be used advantageously in terms of nanoparticle uptake into and within the tumour mass as it permits suitable retention of a therapeutic nanoparticle (TNP). Coarse-structured or 'leaky' tissues of this type result from flawed cancer-tissue vasculature and lead to capillary endothelia *fenestrae* (Figure 7.2). The intraepithelial or transepithelial gap is also different to that in normal tissue, at approximately 200–1200 nm (normal epithelia: 50–100 nm), allowing an enhanced permeability and retention (electron paramagnetic resonance, EPR) effect for TNPs to operate (Table 7.3). The exception to this rule is the organs of the BBB, where the tight junctions may be considerably smaller (at 2–15 nm; Hu and Dahl, 1999; Seelig, 2007; Maeda *et al.*, 2009) than the 20–50 nm seen in most tissues (Davis *et al.*, 2008; Young and Michelson, 2012), and this is important for passive paracellular drug delivery.

Biodistribution of colloidal gold (Sonavanea *et al.*, 2008) and ^{99m}Tc-liposome nanoparticles have been used to probe the optimal penetration size for nanoparticle delivery, which is particularly important in imaging (Wiessleder *et al.*, 1994; Boerman *et al.*, 1997; Sonavanea *et al.*, 2008) and in chemotherapeutic treatment of cancers. The effect of particle size is hugely significant, since extravasation and an increased vascularisation occur as part of the tumour. However, with vascular

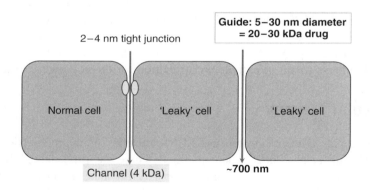

Figure 7.2 The cancer cell: constitution of a tumour and anomalous structure. The figure shows the 'leaky' fenestrated tissue of a neoplastic malignant tumour tissue. Intracell spacing is ∼100× greater than in normal cellular tissues, at 0.6–0.7 μm. Ideal nanoparticle use for inclusion in the tumour by the electron paramagnetic resonance (EPR) effect is then 100–200 nm

Table 7.3 Approved liposome cancer chemotherapeutics

Approval	Trade name	Active	Date approved	Patented	Use
FDA	Taxol	Paclitaxel	1992		CA: lung, ovarian, BRCA, Kaposi
FDA	Doxil	Doxorubicin	1995		CA: wide-ranging, ovarian
FDA	Depocyt	Cytarabine	1999	Yes	CA: lymphoma
FDA	AmBisome	AmphotericinB	1997	Yes	Systemic infection
FDA	Abelcet	AmphotericinB	1995		Systemic infection
FDA	Amphotec	AmphotericinB	1995		Systemic infection
FDA	DepoDur	Morphine	2004	Yes	Pain relief (surgery)
FDA	DaunoXome	Daunorubicin	1996		CA: leukaemia, Kaposi
FDA	Halavan	Paclitaxel	2010	Yes	CA: metastatic BRCA
FDA	Exparel	Bupivacaine	2011/12		Pain relief
EMEA/CPMP (EU/Canada)	Myocet	Doxorubicin	2000		CA: wide-ranging
FDA		Docetaxel liposome	2011		BRCA
FDA	Epaxal	Hep B vaccine	2008		Hep B
	Lipoplatin	Cisplatin	2010	Phase III	CA: ovarian/testicular
FDA		Fluorouracil	2012		BRCA
		Cyclophosphamide + doxorubicin	2012	2004 Phase III	BRCA
FDA	Inflexal V	Influenza IRIV vaccine	2008		Flu
	Gemzar	Gemcitabine	?	2012 Phase II	CA: endocrine
FDA	Marqibo	Vincristine	2012		Rare leukaemia
FDA	Visudyne	Verteporfin	2000		Photo-dynamic therapy
EMEA		Ciprofloxacin	2000 (orphan)		Cystic fibrosis
		SiRNA (plasmid)		2012 CT	Cystic fibrosis
		Isotope, e.g. ^{111}In, ^{125}I, ^{99m}Tc, ^{186}Re, ^{64}Cu (β^+; PET), ^{177}Lu, ^{225}Ac			Brachytherapy, tumour imaging CT, PET, therapy

CA, cancer; β^+, positron emission used in positron emission tomography (PET); CT, computed tomography. 'Phase' in this instance refers to the phase of the clinical study (clinical trials) of drug product development.

endothelia the range of targeting options (Sonavanea *et al.*, 2008) is as follows:

- Most vasculatures are permeable to <10 nm-diameter nanosolids.

- At 15–50 nm, particles (including SLNs) are able to pass through BBB structures.

- Particles as large as 50–200 nm are able to permeate the tumour capillary.

- Particles >200 nm show very restricted permeation (Sonavanea *et al.*, 2008).

Given the chaotic structure of the tumour (the fenestrae (porous tumour windows) can be as large as 1200 nm but are usually ∼300–700 nm (average 500 nm)), there is an optimal size for delivery of approximately 100 nm (Porter *et al.*, 1992; Torchilin, 2001; Moghimi *et al.*, 2005; Al-Hanbali *et al.*, 2006).

For cytotoxics, TNPs and antineoplastics, formulating a drug vehicle at the correct diameter is important because 200 nm particles are more easily retained in the sinuses (Torchilin, 2001; Moghimi *et al.*, 2005) of the liver (Kupffer cell), spleen and lymph nodes, inhibiting prolonged circulatory persistence (Porter *et al.*, 1992; Al-Hanbali *et al.*, 2006) in the blood compartment (Figure 7.3a). Persistence is needed as this reduces the required potency of the particle and its toxicology when administered in 'passive' rather than 'targeted' mode, thus aiding therapy (Valtcheva-Sarker *et al.*, 2007; Vauthier and Couvrer, 2007). At very large nanoparticle sizes (e.g. 500 nm), mucosal ejection is observed, and at >5000 nm, fatal emboli and blood clots can form in the tissues of the lung and brain.

Significantly for drug product designers, 100–180 nm particles are not 'seen' particularly well by the MPS (or RES). Consequently, polymer micelles and SLNs at 10–100 nm are subsumed easily into the tumour (and stay there, exerting a toxic effect) because of its extravasation and saturation of tissues with blood, but are not readily ejected due to the EPR effect (Torchilin, 2001). Much interest has focussed on emulsion, liposome and nanoparticle medicine in the last few years (Muller *et al.*, 1996; Hayes *et al.*, 2006; Jain *et al.*, 2010; Jordan *et al.*, 2012; Lawrence and Rees, 2012). The 'limited' list of approved liposomal cancer therapeutics used routinely in chemotherapy is given in Table 7.3. Therapeutic vehicles and molecular entities have restrictions placed on them by their size and on their delivery by the MPS (shown in Figure 7.3a). In general, large nanoparticles fail because of restrictions in their passage into and within tissues and within the subcellular structures of tissues (Figure 7.3b,c)

7.4 Antimicrobials

Antiinfective (biostatic or biotoxic) drugs may be formulated as topical or systemic DDS dosages. They routinely use incorporation of lethal or semilethal drug

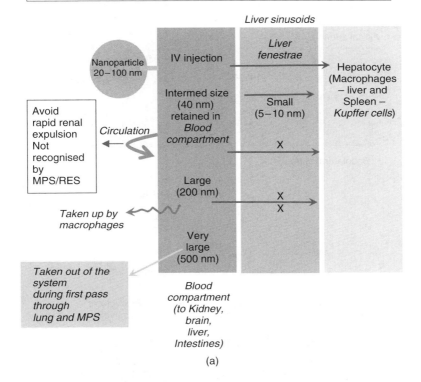

1. Transfer mechanism: fusion, phagocytotic transfer
2. Uptake failure: 7–10 nm trapped in lungs, 500 nm capillary occlusion
Macrophages: nanobead → phagocyte → (lysosome/degradation) → cytoplasm → exocytosis

Liver sinusoids

Nanoparticle 20–100 nm

IV injection

Liver fenestrae

Hepatocyte (Macrophages – liver and Spleen – Kupffer cells)

Avoid rapid renal expulsion Not recognised by MPS/RES

Circulation

Intermed size (40 nm) retained in *Blood compartment*

Small (5–10 nm)

X

Large (200 nm)

X
X

Taken up by macrophages

Very large (500 nm)

Taken out of the system during first pass through lung and MPS

Blood compartment (to Kidney, brain, liver, Intestines)

(a)

Figure 7.3 (a) Circulation of nanoparticles in the blood, including persistence and expulsion processes. (b) Epithelial structure and drug delivery. (c) Mechanisms of uptake involving individual cells, via pinocytosis, phagocytosis or 'active' recognition and uptake, with all three processes ending up with endosomal containment. See Valtcheva-Sarker *et al.* (2007) for fuller details

molecules in an apolar vehicle to disable the infective agent. The most common forms of delivery system include emulsions (Huailiang *et al.*, 2001; Cavalli *et al.*, 2002; Henzl, 2005; Souto and Muller, 2005), SLNs, microemulsions, micelles and liposomes (Table 7.3). Antifungal and antibacterial drugs frequently incorporate fluconazole, miconazole, itraconzole, polymyxins, doxycycline, amphotericin B, penicillin V, cyclosporin, cephalosporins, erythromycin and sulphonamides. Antiparasitic drugs commonly include albendazole and antischistosomal drugs (Araujo *et al.*, 2007). Antiviral pharmaceuticals and emulsions are often fabricated with acyclovir, ritonavir, oseltamivir, zanamivir, tenofovir, darunavir, saquinavir and dapivirine (Grammen *et al.*, 2012).

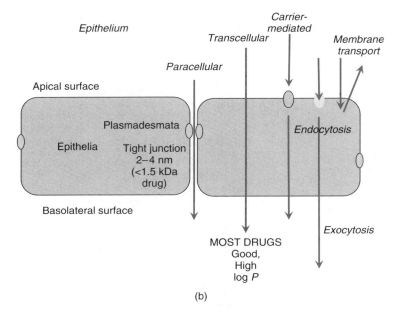

(b)

3 basic mechanisms for a drug nanoparticle (⬤)

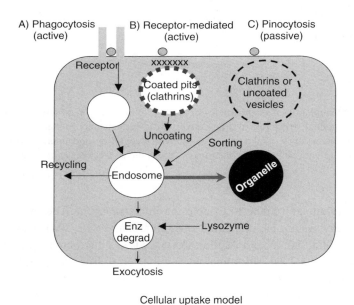

Cellular uptake model

(c)

Figure 7.3   (*continued*)

7.5 Temperature-sensitive matrices and release forms

Block copolymer molecules such as poloxamer 408, along with other polymeric surfactants, can be formulated to show a range of temperature-responsive viscosity changes (Sarker, 2012a). These are used in matrices and release forms where melting (phase change, T_m and liquid-crystal form) and softening can be used for selective application. For polymers such as poloxamer 407 (Pluronic F127), administration in a topical, oral, rectal, ocular or vaginal location can be used to selectively retain and release a 'payload' of drug (Sarker, 2012a). Pluronic F127 is a thermoreversible gelling agent that has been used with cancer drugs such as 5-flurouracil and adriamycin or NSAID analgesics such as indomethacin. Much of their temperature sensitivity and responsiveness is attributed to the lower critical solution temperature (LCST) or higher critical solution temperature (HCST) transition of the palisade layer of 'emulsions' (Figure 2.5) and to the occurrence of pancake (flat to surface) to brush (upright) transformations of surface polymers (Figure 7.4; Al-Hanbali *et al.*, 2006; Concannon *et al.*, 2010; Sarker, 2012a).

This type of behaviour is seen routinely with entangled poloxamers and other block copolymers (F127, PEG, acrylates and methylcelluloses). Polymeric nanospheres (e.g. SLNs and lipid nanocapsules, LNCs) use these inert biocompatible polymers because of their biopharmaceutic properties, such as moderate pK_a, good solubility and solvation, chemical inertness, swelling potential (more

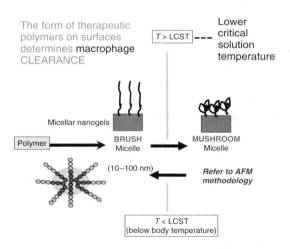

Opsonins (and related plasma proteins) adsorb and label macrophages, then clear.

Figure 7.4 Thermal transitions in the adsorbed polymer. This has implications for stealthification and the creation of long-circulating nanoparticles. LCST is the lower critical solution temperature at which the polymer conformation changes

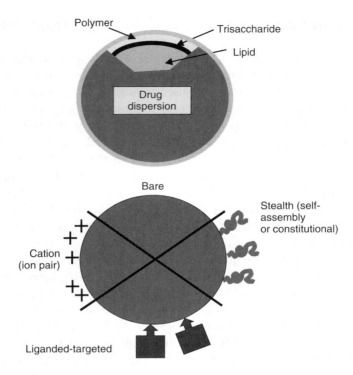

Figure 7.5 Lipid nanoparticles and drug delivery via uncoated species, cation ion pairing, surface antibody conjugation and stealth polymer coating

than a hundredfold), viscosifying capability and modifiable lipophilicity. Other than polysorbate and cetomacrogol emulsifiers, Carbowax (PEG) species are the most widely used water-soluble linear polymers in pharmacy formulation, formed by the addition reaction of ethylene oxide. The simple (therefore favourable) generalised formula based on repeating (n) oxyethylene groups for PEG is:

$$H-(OCH_2CH_2)_n-OH \qquad (7.1)$$

PEG use with molecular weights of 200 Da or more is frequently linked to solution texturisation and stealthification (disguise) of the DDS (Figure 7.5).

7.6 Targeted endosomal use

The cellular activity of drug delivered to a tissue can be made more selective by using prodrugs or antibody attachment (Sarker, 2006a) with the 'cleaveable' conjugation chemistry of disulphide linkage, ester linkage, ether linkage or pH, redox and enzyme (hydrolase, lipase) sensitivity. Examples of this mechanistic approach to drug delivery might include:

- An internal transfer mechanism, driving fusion and phagocytotic transfer.
- Uptake failure resulting from very small or very large particle sizes (7–10 nm trapped in lung, 500 nm trapped in the capillary by occlusion).

Once in the cell, if macrophages engulf the nanoparticle (e.g. liposome, microemulsion or SLN), the resulting loss (refer to Figure 7.3b and particularly c) of the drug (nanoparticle → phagocyte → (lysosome/degradation) → cytoplasm → exocytosis) is seen if it is inappropriately 'disguised' or 'stealthified'. The nanoparticle is internalised normally by pinocytosis, phagocytosis, transcytosis and receptor-mediated active uptake (Valtcheva-Sarker *et al.*, 2007; Sarker, 2012a; Lemke *et al.*, 2013).

7.7 Solid lipid nanoparticles

An SLN is a solid spheroidal lipid–fat–oil particle (Müller *et al.*, 1995, 2000; Müller and Böhm, 1998; Mei and Wu, 2005) which may or may not contain polymer (see Figure 7.5). The similar LNCs have a softer or liquid core and a harder or crystalline shell, as seen with podophyllotoxin (Chen *et al.*, 2006). The surface of an SLN (or LNC) can be smooth or covered by entangled chains (e.g. PEG). An SLN is most characteristically constituted from polymerised lipids and long-chain fatty acids, with or without intercalated polymer, antibody or emulsifier. Some examples of SLN commercial products are given in Table 7.4.

Table 7.4 The diverse range of commercial and approved solid lipid nanoparticle (SLN) drug delivery systems (DDSs)

Name	Type	Dosed in	Use
Duragesic (Fentanyl)	SLN	TD patch	Analgesia
Depocyt (cytosine arabinoside)	SLN (lipid depot)	Cerebrospinal fluid (ISp injection)	Carcinomatous meningitis
Nicoderm (Nicotine) [NiQuitin (Nicotine) Nicorette (Nicotine)]	SLN	TD patch	Smoking cessation
Cyctotoxics (paclitaxel, doxo- and daunorubicin)	SLN	*in aquo*	Parenteral chemo Clinical trials
Qutenza (capsaicin)	SLN	TD patch	Neuralgia

Originators: Gasco (1993), Müller and Lucks (1996).

Considerable 'hot' contemporary research (Müller *et al.*, 2000; Mehnart and Mader, 2001; Cavalli *et al.*, 2002; Rudolph *et al.*, 2004; Souto and Muller, 2005; Kaur *et al.*, 2008; Bhalekar *et al.*, 2009) is looking at the role of SLNs in selective solubilisation of drug in lipid, controlled delivery, various administration routes and future therapeutic perspectives (Üner and Yener, 2007). The loading efficiency of a drug in a lipid is dictated by:

- Drug solubility in the lipid composite (higher solubility means higher loading).

- Drug miscibility in the lipid composite.

- Solid matrix chemistry.

- The polymorphic state of the lipid material (see Section 2.1.2).

SLNs are generally superior to coarse emulsions and nanoemulsions, as the latter suffer from problems of creaming or sedimentation, which lead to coalescence but also to flocculation. Compared to conventional emulsions, SLNs do not suffer to the same extent from problems of poor release control and poor *in vivo* tolerance (parenteral route).

The process of making SLNs was initiated by Speiser in 1990 in a European patent (EP0167825), but it initially resulted in larger sizes and was postulated for crude mass manufacture. The concept developed successfully by Gasco (1993) and Müller and Lucks (1996) was based on a parenteral nanoemulsion formulation from Kabi (Pharmacia), called Intralipid (Moses *et al.*, 2003). Most current forms of SLN use triglyceride as the base (e.g. tristearin), with different modifications of potentially different lipid mixtures, so that it is possible to chemically change the surface to aid in drug delivery (see Figure 7.5). SLNs and core–shell particles (LNCs) can be coated with PEG to prevent blood opsonisation through complement system activation (by providing a dense water sheath) and macrophage clearance (Moghimi *et al.*, 2005).

Control of drug release from the nanoparticle is governed by the drug's size and:

- Diffusion constant (D_{drug}) and interphase diffusion kinetics (k_1, k_2, k_α, etc.).

- Parameters of the Stokes–Einstein equation (see Section 13.2).

- Nanoparticle phase behaviour and thus permeability (T_m, T_g).

- Particle porosity, hardness and polarity (higher log P retained in the fat).

- Enantiotropic or monotropic polymorphism in the fats used.

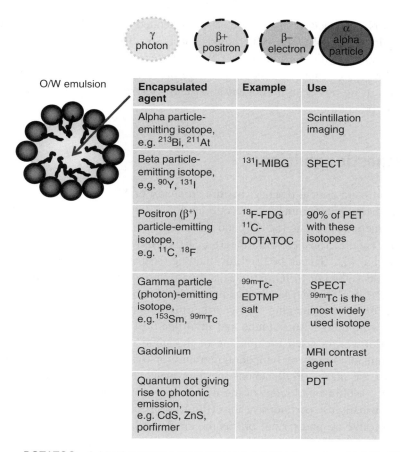

Encapsulated agent	Example	Use
Alpha particle-emitting isotope, e.g. ^{213}Bi, ^{211}At		Scintillation imaging
Beta particle-emitting isotope, e.g. ^{90}Y, ^{131}I	^{131}I-MIBG	SPECT
Positron (β^+) particle-emitting isotope, e.g. ^{11}C, ^{18}F	^{18}F-FDG ^{11}C-DOTATOC	90% of PET with these isotopes
Gamma particle (photon)-emitting isotope, e.g. ^{153}Sm, ^{99m}Tc	^{99m}Tc-EDTMP salt	SPECT ^{99m}Tc is the most widely used isotope
Gadolinium		MRI contrast agent
Quantum dot giving rise to photonic emission, e.g. CdS, ZnS, porfirmer		PDT

DOTATOC = 1,4,7,10-tetraazacyclododecane-1,4,7,10-tetraacetic acid-d-Phe(1)-Tyr(3)-octreotide; MIBG = meta-iodobenzylguanidine; FDG = fluorodeoxyglucose; EDTMP = ethylenediamine tetra(methylene phosphonic acid)

Figure 7.6 Use of emulsions for biomedical imaging. In most cases the identifying species is encapsulated in the emulsion core. Radioisotopes constitute a large majority, including isotopes of iodine, fluorine, yttrium, bismuth, astatine and technetium (metastable), along with optical contrast agents such as gadolinium. The most common forms of imaging are magnetic resonance imaging (MRI), single photon emission computed tomography (SPECT) and positron emission tomography (PET)

Tristearin fats, generally the most frequently used, show monotropic polymorphic change (α (hexagonal) $\rightarrow$ β prime (orthorhombic) $\rightarrow$ β (triclinic)). β-forms of fats have a lower number of defects (therefore expulsion of drug is better), show the highest T_m, have the most stable (chair) conformation and provide the largest crystals (100 μm) compared to the 5 μm α-forms. Phase behaviour therefore influences particle stability and subsequent biological behaviour. A bioactive entrapped within an SLN for injection is customarily suspended in a buffer solution at close to physiological pH (7.4 ± 0.05, blood) and is sterilised by aseptic filtration under clean room conditions rather than thermal sterilisation, but may also be gamma irradiated. Polymerised nanoparticles are prepared by dispersion in multiple emulsions using powerful homogenisation and sonication ($4-6\,°C$), followed by a cleaning procedure.

7.8 Diagnostic emulsions

Emulsions that have optical opacity (barium sulphate) or MRI contrast capability (gadolinium diethylenetriaminepenta acetic acid (Gd-DPTA; gadopentetic acid); superparamagnetic iron oxide (SPIO)) and those which emit measurable positrons (β^+ particles, antielectrons, e.g. fluorine-18 (^{18}F)) or show modification of photon passage (fluorine-19, ^{19}F) can be used for imaging (see Figure 5.2). Incorporation of quantum dots (QDs) or SPIO particles and particles containing gold nanoparticles (Au-TNPs) that respond to illumination or irradiation in generating heat can also be used for thermal imaging (or hyperthermic treatment); these include such products as Cytimmume, Aurimmune and Invitrogen. Ethiodised poppy seed oil, such as lipiodol, can be used for imaging of the small intestine (also using Gd-DPTA oil emulsions) and for lymphanigiography or hysterosalpingography (see Figure 5.2) of the uterus. Perflenapent (perfluoropentane) emulsion is used as a contrast agent for ultrasound scans of the liver, kidneys and vasculature. Diagnostic emulsions usually use perfluorocarbons (PFCs) or naturally occurring vegetable oils (corn oil, olive oil and peanut oil) as the emulsion base (Figure 7.6).

8

Transdermal patches: semisolids

Transdermal (TD) delivery of drugs is influenced by many parameters. Two of the most important with regards to tissue penetration via the stratum corneum (SC) are lipophilicity and molecular weight. For effective delivery, the molecular weights are restricted to <500 Da. The drugs that are delivered routinely by TD routes (see Table 8.1) include:

- oestradiol (296 Da, log P is 4.5);

- scopolamine (303 Da, log P is 0.8);

- clonidine (230 Da, log P is 1.4);

- testosterone (288 Da, log P is 3.8);

- rotigotine for Parkinson's disease (316 Da, log P is 5.0);

- fentanyl (337 Da, log P is 3.9);

- methylphenidate (233 Da, log P is 2.6).

All these species have a characteristically low molecular mass and a moderate to high lipophilicity (log P). TD patches are engineered to have variable content and release profiles but the delivery dose for many therapeutics is usually around 25 µg–20 mg/day.

Most systems or patches fall into two general types: the reservoir form and the matrix (or drug-in-sticky matrix) system. Commercial TD patches include Nicotinell (nicotine in acrylate) and Estraderm (oestradiol in poly(isobytylene)) from Novartis (see Sections 8.1–8.4). The drug is usually dispersed in liquid excipients (i.e. inactive compounds of the vehicle or base). The insoluble porous matrix is usually composed of polypropylene as the main portion with an ethylene vinyl alcohol, polyester or polyurethane backing film and a protective cover membrane. Testosterone (1% w/v) TD patches, for example, can use poly(ethylene glycol) (PEG; the principal ingredient) and water, gelled with methacrylate, as the solvent, but they may also include paraffin oil, silicone oil or terpene oil (limonene,

Pharmaceutical Emulsions: A Drug Developer's Toolbag, First Edition. Dipak K. Sarker.
© 2013 John Wiley & Sons, Ltd. Published 2013 by John Wiley & Sons, Ltd.

Table 8.1   Common drugs delivered by transdermal (TD) technology

Solubilisation of API	Drug/ therapeutic	Commercial patch examples	Use
Nanoemulsion set in polymer matrix	Nicotine	Nicoderm CQ	Smoking cessation
Nanoemulsion set in polymer matrix	Oestradiol	Estraderm and Estrasorb	Hormone replacement therapy (HRT)
Solid nanoemulsion set in polymer matrix	Capsaicin	Qutenza	Neuralgic analgesia: pain relief
Nanoemulsion set in polymer matrix	Testosterone	Androderm	Hypogonadism
Nanoemulsion set in polymer matrix	Ethinyl oestrogen + norethindrone	Ortho Evra and Evra	Female contraceptive
Polymer micelle/ microemulsion	Glipizide[a]	Currently in development and being researched (Glucotrol XL (Actine, Antidiab) tablets exist)	Type II diabetes
Nanoemulsion set in polymer matrix	Scopolamine[b]	Transderm Scop	Motion sickness
Nanoemulsion/ polymer hybrid	Desmethyl diazepam	TDS-Diazepam	Mucoadhesive oral patch. Anxiety and seizure
Polymer matrix suspension	Nitroglycerine	Transderm Nitro, NitroDur and Nitrodisc	Angina
Polymer matrix suspension	Fentanyl[c]	Duragesic	Analgesia (cancer): pain relief; opioid narcotic drug
Nanoemulsion set in polymer matrix	Buprenorphine	BuTrans	Analgesia: pain relief
Nanoemulsion of silicone oil set in polymer matrix	Rivastigmine	Exelon	Alzheimer's disease
Nanoemulsion set in polymer matrix	Oxybutynin	Oxytrol TD	Bladder control/incontinence
Nanoemulsion set in polymer matrix	Diclofenac epolamine	Flector	Analgesia: pain relief
Nanoemulsion set in polymer matrix	Selegiline	Ensam	Anxiety and depression

(*continued overleaf*)

Table 8.1 (*continued*)

Solubilisation of API	Drug/ therapeutic	Commercial patch examples	Use
Polymer in silicone matrix suspension	Methylphenidate	Daytrana	Attention deficit hyperactivity disorder (ADHD)
Nanoemulsion set in polymer matrix	Clonidine	Catapres TTS	Hypertension
Polymer (modified cellulose) matrix suspension	Prilocaine– lidocaine (eutectic mixture)	EMLA	Local anaesthesia prevaccination/ catheterisation
Micelle/microemulsion set in composite matrix	Lidocaine	Lidoderm	Local anaesthesia

[a]Currently in development (e.g. United States Patent Application 20030224050).
[b]1979/1981, first patch approved by the US Food and Drug Administration (FDA).
[c]2005/2007, FDA issues a warning about fentanyl patch safety concerning possible accidental overdose.

eugenol, menthone or terpineol) in order to produce a complex amalgam. TD patches have until recently been most widely associated with oestradiol, testosterone, scopolamine and nitroglycerine delivery (Cevc and Vierl, 2010; Benson and Watkinson, 2012). TD delivery technology can also be linked to electrical and other forms of *in situ* modification, which encourage drug penetration (e.g. iontophoresis, electroporation and sonophoresis) through the skin.

TD technology has the advantages of augmenting bioavailability in bypassing the first pass in the liver and circumventing the gastrointestinal (GI) tract. Careful control of the reservoir (solvent, emulsification, Pickering emulsion) composition can also be used to modify the drug flux (egress) out of the device. Use of cosolvent and penetration enhancer (PEG, PEG-lipids) can further influence the reservoir performance. Flocculation of dispersed nanoparticles in the reservoir occurs when the ζ-potential is lower than $\pm 40\,\mathrm{mV}$. Charging of the particles is made possible by the use of cationic lipids and emulsifier surfactants such as stearylamine ($\mathrm{p}K_a\,11$). Other key formulation ingredients include fatty acids, phospholipids (poloxamers) and an antioxidant (e.g. α-tocopherol).

Passage of the drug transcellularly (through the cells) or paracellularly (between the cells) follows the process shown in the 'brick-wall model' (Moghimi, 1996), represented in summary form in Figure 8.1. The transcellular passage favours hydrophiles, while the intercellular route using the peripheral endogenous lipid phase favours hydrophobes. The model accounts for the complex structural motifs and chemical composition of the skin.

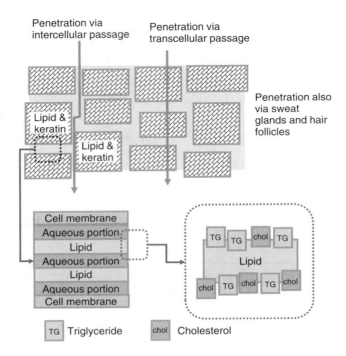

Figure 8.1 The 'brick-wall model' of the outer layer of the human skin and the stratum corneum (SC). The model is discussed at fuller length by Moghimi (1996)

The uppermost layer of the skin is the SC (10–15 cells deep and ~12 μm), which is mainly ~20% lipid, ~20% triglyceride, fatty acids, cholesterol and phospholipids, ~40% β-keratin and ~40% water. Below this, the 2–3 mm sublayer (dermis) consists of ~75% collagen-like material, along with subcutaneous fat and vasculature. Beneath this layer is the muscle base. The rate and extent of drug delivery across the SC is entirely dependent on:

- *In situ* modification of the drug by skin enzymes and microflora.

- Absorption facilitation.

- Solvent type (oil base, water-in-oil (W/O) or oil-in-water (O/W) emulsion).

- Chemical properties of all the strata encountered (hydration, temperature, pH, drug concentration, penetrant and intrinsic skin variations); this is one of the drawbacks to routine use of this technology and its precise modelling.

After an initial slow retarding phase, the concentration of the drug reaches a critical concentration at the interface between the skin and the TD device (Figure 8.2), and so diffusion (uptake of the drug) follows a steady-state process and can be

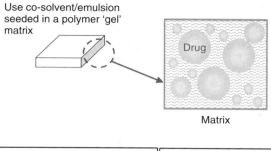

Use co-solvent/emulsion seeded in a polymer 'gel' matrix

Drug

Matrix

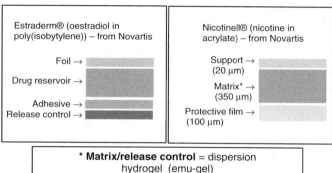

Estraderm® (oestradiol in poly(isobytylene)) – from Novartis

Foil →
Drug reservoir →
Adhesive →
Release control →

Nicotinell® (nicotine in acrylate) – from Novartis

Support →
(20 μm)
Matrix* →
(350 μm)
Protective film →
(100 μm)

* Matrix/release control = dispersion hydrogel (emu-gel)

Figure 8.2 Duo of transdermal patch technologies, showing their use of emulsification and dispersion of apolar drugs

modelled by Fick's law. According to Fick's law, the patch thickness and the drug flux, J, driven by a concentration gradient, are the most significant features of TD delivery (Figure 8.2) and its control. The steady-state process, ignoring the initial equilibrium-forming lag phase, is defined by:

$$J = \frac{DP.(\Delta C)}{\delta} \tag{8.1}$$

$$\frac{DP}{\delta} = kp \tag{8.2}$$

$$Q_t = J.t \tag{8.3}$$

$$Q_t = \frac{(DP\Delta C)(t - \tau)}{\delta}. \tag{8.4}$$

where J is the drug flux (mol/m^2/s), ΔC is the concentration difference ($C_{\text{vehicle}} - C_{\text{skin}}$), D is the drug diffusion coefficient (typically $10^{-12} - 10^{-17}$ m^2/s), δ is the thickness of the barrier (in m), τ is the lag time (s), t is the time (s), P is the solute partition coefficient (oil : water), k is a constant, p is the permeability constant (m/s) and Q_t is the total amount of drug absorbed at time t. The quantity t has been shown to equal ($\delta^2/6D$) and to range from minutes to fractions of a week. In the protein-rich and lipid-rich portions of the skin, the diffusion coefficient (D) and permeability (kp) become significant in determining overall D, which is influenced by moiety size, polarity, shape and charge.

8.1 Hormones

Human growth hormone, like many polypeptides, may be delivered as a nanoparticle using protective polymers and emulsifiers. Oestradiol, testosterone, progesterone and pitocin (syntocinon/oxytocin, often as a pessary) all tend to be delivered in 'oily' vehicles, due to their high log P. The most common routes of pharmaceutical delivery include:

- Parenterals for emulsion-type formulations.

- Pulmonary/nasal in the form of a metered-dose inhaler (MDI) (Videira *et al.*, 2002; Bivas-Benita *et al.*, 2004; Gelperina *et al.*, 2005; Kumar *et al.*, 2008; Mansour *et al.*, 2009).

- Topical creams, lotions and ointments are the most common routes of all for these active pharmaceutical ingredients (APIs) (Cevc and Vierl, 2010; Jones and Harmanli, 2010).

- TD/matrix/implant forms (Cevc and Vierl, 2010; Benson and Watkinson, 2012).

8.2 Analgesia

A range of dosage forms for pain relief, from intravenous (IV) injections to creams (lidocaine) and ointments (ibuprofen), exist as essential therapeutic pharmaceutical emulsions; these also include TD technologies (e.g. nonsteroidal antiinflammatory drugs (NSAIDs), indomethacin, Duragesic, fentanyl, Ropion (flurbiprofen axetil), capsaicin (Qutenza SLN)) and suppositories (paracetamol). Drugs ranging from morphine (and nalbuphine) to prilocaine also use emulsion formulations (Conzen, 2005; Robieux *et al.*, 1991; Fang *et al.*, 2004; Wang *et al.*, 2006, 2008).

8.3 Anaesthesia

Parenteral and topical anaesthesia frequently uses active encapsulation (Almeida and Souto, 2007) in an emulsion or emulsion-bearing organo- or hydrogel. Strategically important products in medicine might include 1% w/v propofol (Diprivan), based on the original Intralipid nanoemulsion formula and used in premedication prior to surgery and general anaesthesia, as well as halogenated emulsifiers (e.g. ~5% w/v halothane/desflurane/sevoflurane (Jee *et al.*, 2012)/isoflurane mixed (Conzen, 2005) with the Liposyn III emulsion system. Local anaesthetics for application as topical creams include the drugs lidocaine (lignocaine), tetracaine, novacaine, fentanyl, pethidine, propofol and etomidate. O/W emulsions are used to solubilise drugs such as clonidine (benzodiazepine class), sodium thiopental and midazolam for rectal delivery in a suppository. Products known as 'rescue emulsions' may also be used to solubilise active, effecting a general anaesthetic removal from the blood, where there are concerns over excessive exposure to toxic levels and overdosing quantities of fluorinated anaesthetics (Lui and Chow, 2010; Ruan *et al.*, 2012; Jee *et al.*, 2012).

8.4 Nicotine

Nicotine is usually taken in a TD adhesive technology/TD patch, such as Nicorette/Nicotinell/Habitrol (Novartis), NiQuitin/Nicoderm (GSK), Nicotrol (Pharmacia) or NicAssist (Boots). These were first developed in the 1980s to counter the desire to smoke by providing a controlled, sustained dose of nicotine. As in many other patches, the active is held within an aqueous/emulsion reservoir within a porous plastic matrix.

8.5 Inserts: vaginal rings

Commercial vaginal ring delivery systems, which are often used to support physiological–surgical interventions (such as a rectocele repair), are usually based upon silicone elastomers, which are normally partial dissolving systems, utilising a variety of materials. The future of advancements in vaginal drug delivery may lie in mucoadhesive and bioadhesive 'plugs' or 'tablets' and in microparticles of hydrogels such as poly(acrylic acid) and modified cellulose or carbowaxes (Acartürk, 2009) with embedded light silicone fluid or emulsion. Well known examples include the Estring (Pharmacia), Femring and Nuvaring, which contain oestrogens (see Section 6.3).

9
Gels

These can be lyophobic, with little interaction between the medium and the dispersed particle, or lyophilic, where the particle and the solvent are similar. Gels can also be of the organo- (e.g. paraffin) and hydrogel (aqueous) type. Stacking in a 'house of cards' form can be based on self-assembly of micro- and nanoparticles, which can include emulsion and solid lipid nanoparticle (SLN) particles.

9.1 Micro- and nanogels

Nanobeads are formed from block copolymer aggregates, and gelation occurs via the three mechanisms represented in Figure 9.1 (see also Figures 3.2 and 7.4). Hydrophilic polymers feature chain entanglements, electostatic association or hydrophobic association, which may be mediated by a bifunctional agent (Sarker *et al.*, 1995b). Hydrophilic particles or polymer-coated hydrophobic solids (e.g. nanoshells), such as those formed from the polymers PMMA, PCL, PLGA, PLA or gelatin, which form core and shell particles, can then associate to form supramolecular assemblies (Corveleyn and Remon, 1998; Hansen *et al.*, 2005), including gels (Mei and Wu, 2005) and microgels that are based on nanoparticles. The systems include nanoemulsion and microgel hybrids for drug encapsulation that are tunable and biodegradable, forming a 30–180 μm bead from sodium dodecyl sulphate (SDS) and poly(ethylene glycol) (PEG)-acrylate ($\phi = 0.2$, 400 nm diameter therapeutic nanoparticle (TNP)), which can be a thermogelling, crosslinkable composite with pockets of nanoemulsion hydrophobicity (An *et al.*, 2012; Hegelson *et al.*, 2012) with immense strength (100 kPa module of rigidity).

9.2 Semisolids

These are typified by gels, pastes, suspensions and emulsions, as discussed in Section 2.1.1, Chapter 5 and Chapter 8). Figure 9.1 shows the fundamental mechanisms of gelation or association used to form macroscopic gels and semisolids.

Pharmaceutical Emulsions: A Drug Developer's Toolbag, First Edition. Dipak K. Sarker.
© 2013 John Wiley & Sons, Ltd. Published 2013 by John Wiley & Sons, Ltd.

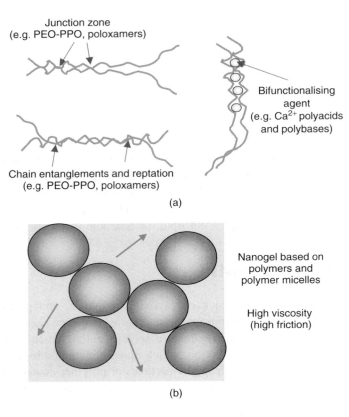

Figure 9.1 (a) The three primary mechanisms behind gelation. (b) The determinants of higher viscosity in a sample are friction and internal fluid/solvent capture and flow

Semisolids (Jones and Harmanli, 2010) are usually coarse dispersions (emulsions, sols, gels) or poorly or 'unstructured' dispersions of water in solids or oils and fats. Typical semisolids include hydrogels, organogels, creams, lotions, ointments, implants, pastilles and topical bases.

10

Implants

Subcutaneous implants usually derive from gels, apatites (hydroxyl calcium phosphate) and solid foams of porous interconnected polymer films or strands formed from nanobeads. Implants have a variety of roles (e.g. hormone release), through permitting continuous dosing; they initially penetrate the skin via an invasive (surgical) procedure. The implant thus acts as a reservoir and source of drugs directly accessible to the body. The most common physical forms of implants and composite materials are represented in Figure 10.1.

Implants can range from autonomous micropumps for analgesics and reservoirs for hormones to implantable drug delivery systems (DDSs; e.g. insulin peptide in a solvent in a porous matrix) or polymer matrices for contraceptives (e.g. Norplant). Fouling of the implant by biofilm formation can be an issue, reducing product efficacy. This is controlled by the inclusion of cell growth inhibitors and their measured leaching out following surgical implantation.

10.1 Plastics and glasses

This technology is suggested for use with dispersed nanosolids (Hansen *et al.*, 2005; Boevski *et al.*, 2011), for example solid lipid nanoparticles (SLNs) of an oligosaccharide such as trehalose, starch or carbopol polymer. In the dry state, the drug is trapped in a glassy core (Corveleyn and Remon, 1998); upon hydration, it is contained in a semigelatinous layer. Continued gelation influences drug efflux differently than a traditional matrix tablet: the thermodynamic activity or chemical potential of the gel matrix increases when the gel layer around the dry tablet increases, the core acting like a rate-controlling membrane, resulting in linear release of the drug. Performance-controlling factors include:

- crosslinking density;
- chain entanglement (see Equation 3.7);
- crystalline phases, making up the composite along with pH and ionic strength.

Pharmaceutical Emulsions: A Drug Developer's Toolbag, First Edition. Dipak K. Sarker.
© 2013 John Wiley & Sons, Ltd. Published 2013 by John Wiley & Sons, Ltd.

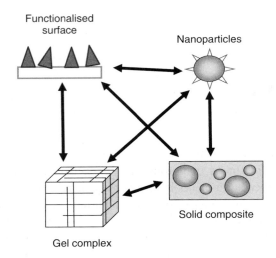

Figure 10.1 Composite materials and their potential uses in drug delivery platforms and innovations

10.2 Thermoresponsive materials

These cover a phenomenal range of biomaterials (Qian *et al.*, 2013; Rao *et al.*, 2013), biomimetic composites (Figure 7.4) and truly novel materials (Sarker, 2010, 2012a). Their responsiveness comes from temperature-sensitive particles, polymers, liquid crystals (Jones, 2002) and nanoparticles, such as microemulsions (Lawrence and Rees, 2012) and niosomes (Kazi *et al.*, 2010), which frequently undergo phase inversion and interconversion as temperatures are modified (Corveleyn and Remon, 1998; Hansen *et al.*, 2005).

Temperature-sensitive polymers (e.g. poly(N-isopropyl acrylamide), pNIPAAm) and chemical derivatives are used as hydrogels (particle dispersions in gel form) for drug delivery, commonly showing gel–sol transitions that are dependent on temperature (Corveleyn and Remon, 1998; Hansen *et al.*, 2005). Changes can be based on the particle droplet surface or polymer coating of the emulsion (dispersion), demonstrating hydrophilic–hydrophobic transitions in response to changes in environmental conditions such as temperature (Fitzpatrick *et al.*, 2012). Temperature-sensitive drug release systems based on phase-change materials (PCMs) can also include colloidal particles. Above the melting point, the PCM melts, permitting the encapsulated oil-in-water (O/W) emulsion particles (i.e. poly(ethylene glycol) (PEG), poloxamer or PLGA covered droplets), vesicles, microemulsions and SLNs (Almeida and Souto, 2007) to leach out. External and internal stimuli that facilitate material change (Boevski *et al.*, 2011) include temperature (but additionally pH, ultrasound) and specific substrates (e.g. Ca^{2+}; Choi *et al.*, 2010).

11

De novo science, sustainable novel products and platform applications

The current most explored areas of drug development include nanoemulsions (solid, liquid and core–shell types), theranostics (complex composite materials), multimodal imaging and cancer nanotherapeutics (Gianella *et al.*, 2011). Multimodal or many-use nanoparticles are at the forefront of 'theranostic' (therapeutic–diagnostic) hybrid systems. The use of theranostic platforms can be based on oil-in-water (O/W) nanoemulsions with superparamagnetic iron oxide (SPIO) nanocrystals, nanosized perfluorocarbon droplets or gadolinium diethylen-etriaminepenta acetic acid (Gd-DPTA) for magnetic resonance imaging (MRI). The fluorophore 'Cy7' is now used for near-infrared (NIR) fluorescence imaging, other photoresponsive materials such as quantum dots (QDs) and nanogold particles are used for hyperthermic treatments (photodynamic therapy, PDT) and the hydrophobic (high-log *P*) glucocorticoid drugs such as prednisolone tagged with valeric acid are used for therapeutic purposes. Targeted nanoemulsions surface functionalised with $\alpha v \beta 3$-specific RGD-amino acid peptides have also been used for specific cellular targeting (Gregoriadis, 1977). Fluorescent O/W nanoparticle platforms of this type can be applied for imaging guided cancer therapy.

In this vein, cellular-responsive mannosylated O/W nanoemulsions carrying a 'payload' of antitumour therapeutic (Davis *et al.*, 2008; Torchilin, 2001) are used for cell-specific drug delivery (Guo *et al.*, 2012). In an interesting 'slant', the tumour-targeting potential of surface-modified solid lipid nanoparticles (SLNs) loaded with doxorubicin has been studied for cancer treatment (Almeida and Souto, 2007). Thus, encapsulating SLNs were mannosylated (Jain *et al.*, 2010) to aid targeting and efficient uptake. Nanoemulsion anticancer therapeutics (Sonavanea *et al.*, 2008) for oral use (Tang *et al.*, 2012) (e.g. analgesia, but more conventionally systemic applications) are a recurrent theme of *de novo* technologies (Jordan *et al.* 2012). Continuing with self-assembled nanoparticles, a survey of microemulsions (Lawrence and Rees, 2012) and their use in drug

Pharmaceutical Emulsions: A Drug Developer's Toolbag, First Edition. Dipak K. Sarker.
© 2013 John Wiley & Sons, Ltd. Published 2013 by John Wiley & Sons, Ltd.

delivery was undertaken recently. Since microemulsions are close to the ideal size for cell, vascular and cytoplasmic permeation at 10–50 nm, these particle colloids, which are clear, thermodynamically stable, isotropic mixtures of oil, water and surfactant/cosurfactant, seem at present to be vastly under-exploited. The ability to incorporate a plethora of drugs makes them of great interest to the drug developer (Lawrence and Rees, 2012).

Among post-2000 technologies (see Figure 11.1), where there are many patents for uses ranging from solubilisation aids to nanogels to hydrogels, one conceptual idea stands out: Popp's innovation (Popp, 2009) describes a composite material (Corveleyn and Remon, 1998; Hansen *et al.*, 2005) made from polymer and nanocrystals (QDs) as a superbly utilitarian potential pharmaceutical dosage-form platform technology (Sarker, 2010). While requiring US Food and Drug Administration (FDA) clearance, this technology is responsive to pH, temperature, pressure, chemical triggers (Boevski *et al.*, 2011), mechanical shear and

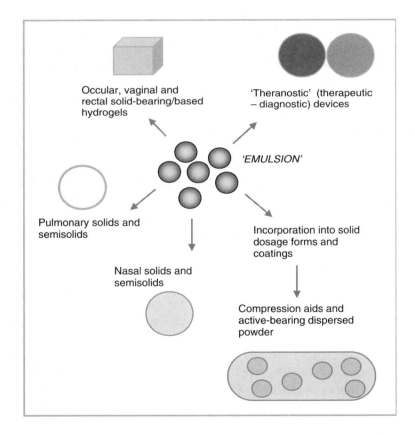

Figure 11.1 The diversity of emulsion and lipidic drug delivery systems (DDSs) and therapeutic–diagnostic (theranostic) devices

redox sensitivity to the environment in producing materials for specialised drug delivery and encapsulation (and self-indicating pharmaceutical packaging). Liquid crystals and specialised niosomes (cubosomes: liquid crystal bicontinuous structured, particle-based nanostructures) show significant potential in pharmacy.

De novo science efforts to produce sustainable novel products have also often centred around fluorinated glucose (^{18}F-FDG) cancer cell metabolism and positron emission tomography (PET) imaging (e.g. breast cancer). Pickering emulsions show great future promise and a lack of sensitivity to temperature (Binks, 2002; Arditty *et al.*, 2004). Solid emulsions have the advantage of offering control over drug release, which is not offered by liposomes, micelles, microemulsions or coarse/fine emulsions (Müller *et al.*, 2000; Sarker, 2010, 2012a).

Recent advancements in drug delivery technologies (see Table 11.1) include:

- Controlled release and self-emulsifying systems.

- PEGylated and 'stealth' therapeutics (Torchilin *et al.*, 2001).

- Inhaled therapeutics.

- Solid and solid-shell emulsions (Almeida and Souto, 2007).

- Transdermal drug delivery technology.

- Implant-based drug delivery.

- Oncology therapeutics (Gianella *et al.*, 2011).

11.1 Tablets

A new wave of product actives (active pharmaceutical ingredients) may be bound in compressed form within SLN-based emulsions (Noguchi *et al.*, 1975; Pouton, 2000; Brusewitz *et al.*, 2007; Sarker, 2012a), which are themselves dispersed in a tablet or a 'glassy' medium of polymer such as HPMC-SDS or maltodextrin (Corveleyn and Remon, 1998; Hansen *et al.*, 2005). This is illustrated in various possibilities for solid dispersions in Figures 10.1 and 11.1. The evolution of this type of product sits right at the cutting edge of SLN use and illustrates the applicability of the 'emulsion'.

11.2 Metered-dose inhalers

Aerosolised solid nanoparticles and emulsions enter the lung, which presents itself as an aqueous vascular space well suited to drug delivery and gene therapy (Bivas-Benita *et al.*, 2004). Metered-dose inhalers (MDIs) are currently being used for

Table 11.1 Current and in-development nanoparticle drug delivery systems (DDSs)

Species	Drug (therapeutic agent)	Therapeutic	Route	Type of soft matter
Polymer micelle	Doxorubicin	Cancer	IV	Poloxamer, poloxamine, polysorbate
	Daunorubicin	Cancer		
	Haloperidol	Antipsychotic		
	Digoxin	Cardiac		
	Indomethacin	Analgesia		
	Ibuprofen	Analgesia		
SLN	Capsaicin	Analgesia	IV, TD, topical	PEG-lipid/ tristearin melt
	Nicotine	Replacement		
	Doxorubicin	Cancer		
	Azidothymidine (AZT)	HIV		
	Lidocaine	Anaesthesia		
Nanoemulsion	Dicarbazine	Cancer	IV, TD, topical	Oils with: span, Polysorbate, PEG, poloxamine, poloxamer, carbomer, lipids
	Propofol, dex-amethasone, acyclovir, saquinavir	Hydrophobe inclusion		
	Nicotine, oestradiol	Replacement, HRT		
	Ritonavir	HIV		

IV, intravenous; SLN, solid lipid nanoparticle; TD, transdermal; PEG, poly(ethylene glycol); HRT, hormone replacement therapy; HIV, human immunodeficiency virus.

peptides such as insulin (Exubera). Future and current research is looking at formulations based on either SLN emulsions (Videira *et al.*, 2002; Mansour *et al.*, 2009) or lipid nanocapsules designed based on a 'melt' with a hydrophilic polymer (e.g. PEG) or soft-conjugated to a soluble carrier (e.g. Pluronic/Tetronic).

11.3 Blood substitutes

Also known as artificial blood/blood surrogates, these are frequently aphron, micellar, nanogel and microemulsions/nanoemulsions based on perfluorocarbon

(Riess and Krafft, 1998; Krafft and Riess, 2009), such as perfluorodecalin and fluorinated surfactant (Courier *et al.*, 2004), or biocompatible amphiphiles, such as human–bovine serum albumin/lecithin. They are used ubiquitously because of the ease with which they solubilise and thus carry oxygen, and are thus suitable for use as universal blood substitutes. Examples (see Figure 5.2) include:

- Oxygent, not currently FDA approved;
- Fluosol DA, FDA approved;
- Perftoran/Perftec, approved Russia/Mexico;
- Oxycyte, in trial;
- Oxyfluor, discontinued;
- Sangart, a PEG-modified haemoglobin.

III

Tests: chemistry to control the quality, efficacy and fitness for purpose of a product

This section is based on a number of key activities used routinely in quality control (QC) to achieve a high degree of assurance of quality. At the most fundamental level, this involves physical sampling and provision of confidence in that sampling, which means reproducible sample recovery, achieved by reliance on valid analytical methods (VAMs). Sampling regimes can be used as an addition to in- and on-process tests (see process analytical technologies, PATs). PATs for liquid emulsion products often take the form of a 'magic eye' based on metal presence, colour and turbidity. Rapid tests account for a large number of PATs, including impedance and epifluorescence for microbial burden and impurity limit tests (chemical analysis) of the sample. For many products, asepticity is a prerequisite for use, but the severity of in-process intervention can influence drug solubility, drug polymorphism and thermocatalytic change.

Expiry date and shelf life are modelled using the content of drug active pharmaceutical ingredient (API), excipient degradation (Sarker, 2002, 2008) and storage conditions (temperature versus humidity and light). Any well-thought-out QC regime will have intercalated check point testing (hazard analysis of critical control points, HACCP). Standard indices for emulsion and colloidal systems are quantifications of the following: drug/impurity assay (usually by reversed-phase high-performance liquid chromatography (RP-HPLC), gas liquid chromatography (GLC) or ultraviolet (UV) spectroscopy), visual appearance, odour (or headspace gas chromatography), rheology/viscosity, opacity, particle size/shape/charge, melting point, water content and biopharmaceutic profiling (Sarker, 2006, 2012a; Al-Hanbali *et al.*, 2006). As part of this profiling, characterisation (also primarily undertaken and identified at the preformulation stage) will need to consider

Pharmaceutical Emulsions: A Drug Developer's Toolbag, First Edition. Dipak K. Sarker.
© 2013 John Wiley & Sons, Ltd. Published 2013 by John Wiley & Sons, Ltd.

drug entrapment efficiency (see Equation 1.1), *in vitro* release (e.g. via standard or bespoke dissolution apparatuses (suppository)), dialysis tubing (nanoparticle) or the Franz diffusion (topical product) cell (at 25 or 37 °C). In some cases, evaluation of form and the suitability or compatibility of excipients and drug may involve histological, tissue/cell culture assay and toxicology studies.

12
Physicochemical properties

Crucial physicochemical properties that define, underpin and dictate therapeutic performance include product pH generally, ranging from mildly acidic to very slightly basic (i.e. pH 3–8), pK_a and thus ionisation (and *in vitro–in vivo* solubility), moiety apolarity or log P (log D) and buffering capacity (β)/resistivity of the stabilising buffer salts and the drug itself. There are also obvious drug or carrier size constraints, and these influence both diffusion and solubility or ease of penetration within tissues. The concentration of hydrogen ions (pH) is a measure of acidity and is defined by:

$$pH = -\log_{10} [H^+] \tag{12.1}$$

where $[H^+]$ is the concentration of protons. For example, the pH of the stomach and of strong mineral acids is $\sim$1.0, whereas that of the skin is $\sim$4.9, the vagina and colon are 4 and $\sim$6.3, respectively, the blood is 7.4 and the lung is 7.4. This regional variability (see Figure 1.1) has connotations for effective drug delivery, as in the case of small lung cell carcinoma (cancer). The tumour intracellular pH and extracellular pH are different, and have been measured at 7.2 and 6.8 respectively. This would be very significant in terms of bioavailability if the tumour-targeted drug had a pK_a of 7 (Gerweck *et al.*, 2006). The pH of the drug and the excipients is important, as it pertains to stability; for example, a 1% w/v aspirin solution has a pH of 3.5, pure olive oil 5.6, pure arachis oil 6.8 and cocoa butter 5.5. Degradation and hydrolysis are often higher in lower and high-pH environments. For an acid species (a potential donor of protons), the dissociation constant is:

$$pK_a = -\log_{10} ([products]/[reactants]) \tag{12.2}$$

Consequently, when the pH $\ll pK_a$, the acid drug or excipient species is largely un-ionised (i.e. a potential H^+ donor), but for acid drugs or where pH $\gg pK_a$, the species is fully ionised. The pK_a can be used to conceptualise the 'strength'

Pharmaceutical Emulsions: A Drug Developer's Toolbag, First Edition. Dipak K. Sarker.
© 2013 John Wiley & Sons, Ltd. Published 2013 by John Wiley & Sons, Ltd.

of an acid or a base. This is demonstrated for example as:

- a pK_a of -7 for hydrochloric acid (strong acid);
- a pK_a of 2.75 for penicillin (acid);
- a pK_a of 2.79 for apolar benzylpenicillin (acid);
- a pK_a of 3.5 for aspirin (acid);
- a pK_a of 4.75 for acetic acid (weak acid);
- a first pK_{a1} of 7.4 for phenobarbital (acid);
- a pK_a of 9.2 for ammonium cation (acid);
- a pK_a of 9.5 for paracetamol (acid).

However, confusion may arise since:

- a pK_a of 3.3 is seen for diazepam (base);
- a pK_a of 9 is seen for diphenhydramine (strong base).

Multiple pK_as are routinely seen with a multitude of drugs:

- amphoterics such as adrenalin (epinephrine) have pK_as of 8.7, 10.2, 12;
- morphine has pK_as of 8, 9;
- pharmaceutical buffering amino acids (e.g. glycine) have pK_as of 2.3, 9.6.

A pK_a above 7 does not necessarily imply a base. However, things are not quite so straightforward in any case, as many polyprotic drugs have multiple pK_as (where the overall value is simply an average). Adrenaline has several pK_as, at 8.7, 10.2 and 12, and hence its overall pK_a may be calculated as $(8.7 + 10.2 + 12)/3 = 10.3$.

When the $pH = pK_a$, the species is 50% ionised, which as shown in Equation 12.3 means ideal buffering capability. Knowing the pK_a is also crucial (Figure 12.1) because it points at the degree of ionisation (see Equation 12.2) and therefore the likely bioavailability. For example, in the blood or lung at pH 7.4, itraconazole ($pK_a = 3.7$), phenytoin ($pK_a = 8.1$), streptomycin ($pK_a = 8.7$), propofol ($pK_a = 11$) and dexamethasone ($pK_a = 13.5$) are likely to be in the first-case un-ionised, with all four others being essentially ionised/partially ionised. Equations 12.1 and 12.2 are amalgamated in the extremely useful Henderson–Hasselbalch equation:

$$pH = pK_a + \log_{10} ([\text{products}]/[\text{reactants}]) \qquad (12.3)$$

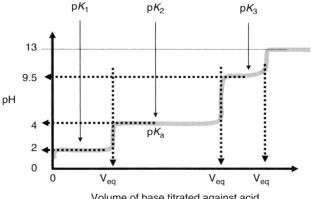

Figure 12.1 Representation of a titration curve for polyions and buffers, and an estimation of ideal buffering (pK_a = pH) and ionisation. The titration curve for a tribasic (triprotic) acid is shown

Another important concept is that of oil (octanol as a model apolar solvent) solubility or log P. This index is extremely helpful, as is pK_a, for understanding the complexities of bioavailability and cellular uptake. For example, the cytotoxic drug doxorubicin ((pK_a = 8.1), LD_{50} = 0.022 g/kg rat (subcutaneous)), has a low log P of 0.9 and a low water solubility of 1.2 g/l; here the log P and pK_a hint at a polar structure, but also how it might best be constituted given that the drug is not easily dissolved in water.

$$\log P = \log_{10} ([\text{drug in octanol}]/[\text{drug in water}]) \tag{12.4}$$

As a 'rule of thumb', when the log P is large and positive, the drug is regarded as more lipid soluble; when log P = 0, the drug is equally soluble in both phases (cf. negative $\sim$ water soluble). Examples of drugs traversing the extreme hydrophile and lipophile are shown in Table 12.1. The extreme lipophilic drugs such as testosterone (3.3) and Terbinafine HCl (6.5) require special solubilisation in oils (Sarker, 2005a, 2012a) and lipidic emulsifier and surfactant-based particles.

Bioavailability and an absorbed fall-off in activity are seen after a log P optimum is reached and can be used to aid formulation. The optimum is frequently caused by a solubility limitation or the extent of unbound drug in the blood plasma. Notably, overtly apolar drugs do not pass in or out of the cell membrane easily and thus have poorer bioavailability.

Accurate predictions of log P are now possible (Table 12.2), and these are very useful in selecting the dosage form in preformulation. Software for carrying out estimates includes a number of fragmentary and atomic models, such as cLog P, Xlog P and Moriguchi Mlog P. Most reference measurements for

Table 12.1 Physicochemical parameters of importance during formulation

Formulation consideration	Amount of drug available (bioavailability)	Physicochemical properties of the drug already understood and of use in providing new information in final product manufacture	Anticipated dose and optimal route of delivery
Assay	✓		
Log *P*/solubility in solvents	✓	✓	✓
pH/pK_a	✓	✓	✓
Chemical change: racemisation, polymerisation, excipients compatibility	✓		
Melting point/ polymorphism		✓	✓
Stability (light, oxygen, thermal)	✓	✓	✓
Flow/rheology			✓
Particle size/form	✓	✓	✓

log *P* are undertaken at 25 °C, but their values are questionable to an extent with gut/blood (37 °C) and skin (33 °C) bioavailability. More hydrophobic drug molecules generally become less soluble as the temperature increases. However, in aqueous media this is complicated by entropic factors, due to their size and/or charge plus specific structural effects. The effect of solvent and pH is factored into the distribution ratio:

$$\log D = \log P - [1 + \text{antilog}(\text{pH} - pK_a)] \tag{12.5}$$

Log *D* is the ratio of the equilibrium concentrations of all species (un-ionised and ionised) of a molecule in octanol to the same species in the water phase at 25 °C. It varies from log *P* in ionised species and the neutral form of the molecule. However, ion pairs of drug can combine, as $A^- + B^+$ gives A-B (neutral), which is more lipid soluble. This is difficult to anticipate. Despite its attempt at prediction or foresight, log *D* tends not to be used as routinely as log *P*.

Water-based solubility centred about the dielectric constant (ϵ) ranges from 80 in water to 50 in glycols, through ~20 in aldehydes and ketones and 5 in alkanes, to ~0 in heavy oils. The lower values are said to be indicative of increasing apolarity, but since some extremely polar molecules (HCl, triethylamine) have values less than 5, the dielectric constant is not an accurate prediction of the 'full story'

Table 12.2 Log P and lipophilicity across a range of drugs used for diverse therapies. Cosolvents such as macrogols (poly(ethylene glycol), PEG) and ethanol-propylene glycol are used to aid solubilisation of log P compounds in the range 2–4

Compound	Log P value	Character	Properties
Oxytetracycline	−1.1	Hydrophile	Poor absorption
Caffeine	0.01	Hydrophile	
Aspirin	0.1	Hydrophile	
Prednisone	1.5	Moderate lipophilicity	Good absorption
Temazepam	2.2	Moderate lipophilicity	Good absorption
Testosterone	3.8	Lipophilic	
Propranolol	3.6	Lipophilic	
Hydrocortisone	4.3	Lipophilic	
Mometosone furoate/ betamethasone	4.7	Lipophilic	
Itraconazole	5.7	Extreme lipophilicity	Difficultly in 'normal' formulation
Terbinafine HCl	6.5	Extreme lipophilicity	Difficultly in 'normal' formulation

Example cosolvents: ethanol/propylene glycol mix – diazepam; macrogols (PEG) – hydrocortisone; microemulsions and micelles – vitamins ADEK, miconazole, indomethacin.

in explaining solubilisation. In this case, moiety polarisation and dipole moment are also relevant. Solubility can also be enhanced by using complex vehicles (in addition to the main solvent). Clobetasol propionate (steroid) shows a 100-fold increase in aqueous solubility when the cosolvent propylene glycol (Busse, 1978) content is increased from 30 to 60%. Similar solubilisation enhancement might be anticipated via solubilisation in dispersed micro- and nanoemulsions.

Le Chatelier's principle describes the effect of an ion on the push of equilibria towards reactants or products. One variant is based on pH and the effect of adding an acid. As shown in Equation 12.6, this simply increases the amount of un-ionised drug, 'HA', available to be dissolved in an apolar phase (e.g. oil in an emulsion), or in some cases drug can be solubilised in the 'amphiphile' layer (Sarker, 2009a) of a vesicle, microemulsion or micelle:

$$[A^-] + [H^+] \rightleftharpoons [HA_{aq}] \rightleftharpoons [HA_{org}] \qquad (12.6)$$

where org represents the organic or aliphatic phase and aq represents the aqueous phase.

Emulsification (Sarker, 2005a, 2006a) is a form of encapsulation that may be used for poorly water-soluble compounds (steroids, antibiotics, vitamins ADE, biomedical imaging agents, lycopenoids, terpenoids) for the purposes of specific delivery and flavour masking, protection or compounding. For encapsulated drugs, Fick's law conveniently describes the key parameters that govern drug release from liquid droplets:

$$(dm/dt) = -DA(dc/dx) \qquad (12.7)$$

where t is the time and m is the mass concerned. The law states that there is a concentration gradient following diffusion (D), moving away from the vehicle; that is, concentration (c) 'gets weaker' at distances (x) further from the particle surface. This is elaborated on by the Noyes–Whitney equation, which goes on to describe the stationary 'boundary layer' of molecules at the particle surface and the importance of surface area (A) for dissolution:

$$dm/dt = \frac{solubility\ constant \times A}{h} \qquad (12.8)$$

where h is the layer thickness. Needless to say, dissolution from a liquid droplet, micelle, nanoemulsion, solid lipid nanoparticle (SLN) or liposome follows the same general process as dissolution of or from a solid matrix.

12.1 Thermal evaluation (differential scanning calorimetry) and lipid polymorphs

A multitude of thermal analysis techniques are used to investigate pharmaceutical materials. They measure a physical property of the material as a function of temperature. The less commonly used include thermomechanical analysis (TMA), dielectric analysis (DEA) and hot-stage microscopy. More frequently useful for pharmaceutical materials are thermogravimetric analysis (TGA; Figure 12.2), differential thermal analysis (DTA) and the temperature-related viscometer. The most useful and arguably the most often used, with regards to polymorph discrimination (Figure 12.3), crystalline to amorphous ratio estimation, purity and assay, is differential scanning calorimetry (DSC; Figure 12.4). DSC is used for examination of semisolids, gel-phase lipids (see Figure 2.3 and Figure 7.1), granular solids and composites where heterogeneity of samples is detrimental, as this influences heat conduction (coefficient of conductivity) and specific heat capacity, ΔC_p (the way in which heat is transferred through a solid). DSC equipment operates over temperatures up to $1600\,^{\circ}\text{C}$, needs only micrograms to milligrams and can be fully automated for 30–100 samples. Thermogram quantitation can discover the latent heat (enthalpy) for a process (see Figures 12.3 and 12.5):

$$A = -kGm\Delta H \qquad (12.9)$$

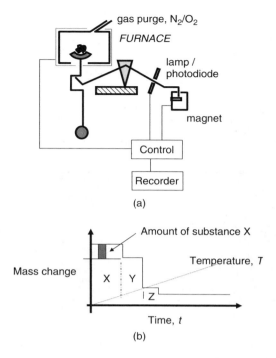

Figure 12.2 (a) Thermogravimetric analysis (TGA) and (b) its use for estimation of content based on thermal reactivity

Using the same equipment reduces the term thus (since kG → 1):

$$\Delta H = A/m \tag{12.10}$$

where A is the peak area, k is the thermal conductivity constant, m is the mass, G is a geometric calibration factor and ΔH is the enthalpy. An endothermic or exothermic process is signalled by an easily quantifiable enthalpy, ΔH. A peak on a thermogram indicates a process is:

- Endothermic (positive ΔH), e.g. melting (see Figure 12.5), vaporisation, sublimation, adsorption, desorption.

- Exothermic (negative ΔH), e.g. crystallisation, oxidation, polymerisation, catalysis.

Peaks on a thermogram indicate a phase change or change of state, such as the π-transition in lamellar phase lipids (Figure 2.6). The glass transition temperature (T_g) is consistent and a characteristic for polymers and amorphous noncrystalline materials with a typical $\Delta H = 0$. This is seen with an amorphous 'glass', where

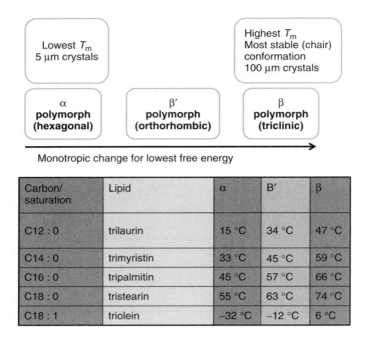

Figure 12.3 Polymorphism in triglycerides. Polymorphism is frequently implicated in variations to crystal size, porosity and melting point

specific heat capacity change is due to segment motion, which is affected by crosslinking, the surroundings and molecular asymmetry and is seen as a 'step' in the thermogram. Melting point (T_m) can be measured and is typically narrow for pure substances (Chaiseri and Dimick, 1986; Freitas and Müller, 1999) and broad or ill-defined for mixtures. The technique is invaluable in any work dealing with polymers (e.g. emulsifiers), self-assembled structures (e.g. liposomes) and vehicle fats or oils. Some useful definitions relating to materials and thermal analysis are:

- *Crystal structure*   Seen as long-range order, regular molecular periodicity. Fixed crystal habit, definite melting points (lattice energy).

- *Amorphous solids*   Created by rapid cooling, no long-range order in solid. Can be considered as super-cooled liquids with random molecular arrangement (a glass).

- *Crystallinity*   The degree to which structural order (determines hardness, porosity, density, melting point, solubility and thus bioavailability) is present.

- *Hygroscopicity*   The ability to attract water molecules, e.g. glycerol, absorption bases and emulsifying waxes (see Chapters 5 and 6). Important in understanding lyophilisation.

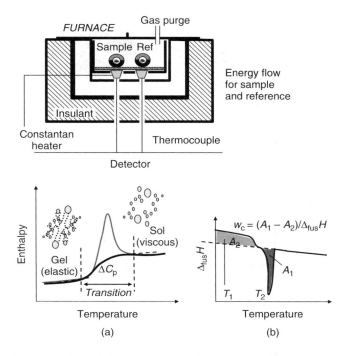

Figure 12.4 Differential scanning calorimetry (DSC) apparatus. (a) Its use for following gel–sol transitions. (b) Determination of the degree of crystallinity (W_C) based on a endothermic melting thermogram

- *Deliquescence* Solubilisation of drug in absorbed solvent/adsorbed water.

- *Polymorphism* A very common variation in form which influences porosity/packing and hygroscopicity. Different physical crystal forms of the same (identical) chemical substance exist, varying in their melting points and solubilities. This crystal form affects dissolution and biological availability.

The lowest free-energy states prevail between metastable forms, such as theobroma oil (Chaiseri and Dimick, 1986). Other examples include Rofecoxib (Vioxx), used for arthritis, which demonstrates five differing morphological characteristics based on crystallisation solvent. Polymorphism is also seen routinely with steroids (cortisone: 2 polymorphs; prednisolone: 2 polymorphs; flucortisone: 2 polymorphs and 19 pseudopolymorphs). Barbiturates (barbitone: 6 polymorphs; pentobarbitone: 3 polymorphs) and antiviral drugs (Ritonavir: 2 polymorphs) also demonstrate this clearly. The range of melting temperature (T_m) variation of polymorphs (drugs and excipients) can be significant, as illustrated by pentobarbitone, with (i) 129 °C, (ii) 114 °C and (iii) 108 °C, and theobroma oil, with ranges of

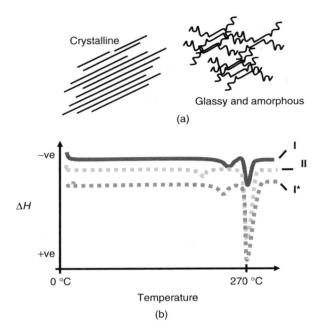

Figure 12.5 (a) Phenomenological model of crystalline (regularity of structure) and glassy or amorphous samples. These would typically be fats and polymeric materials (or two forms of drug actives, e.g. insulin), respectively. (b) DSC thermogram of an anonymous drug, showing the three distinct polymorphs present and the variation in the appearance of the DSC trace (the variants are notionally classified as types I, I* and II)

18–35 °C (see Table 2.5). Pseudopolymorphism is where different crystal shapes result from hydration and solvation of a solid. Enantiotropism is a two-directional change between two (or more) polymorphic states, whereas monotropism is unidirectional change between states, as seen with cocoa butter (Table 2.5). Observed melting or phase changes are higher when:

- The species has a higher lattice energy.

- There is significant chain stiffness.

- There is increased molecular weight.

- The crystallites increase in size.

Lower solubility, due to polymorphism, may then explain differences in bioavailability between batches of drugs and drug products. Drugs such as novobiocin (polyheterocyclic drug) and the peptide insulin have an amorphous form, which is more soluble than the crystalline form; in this case, lower lattice energy lies behind easier solubilisation. Some solid dispersions form simple eutectic mixtures (a mixture having a single chemical composition); for example, lidocaine and prilocaine, both solids at 25 °C, form a eutectic mixture liquid with a T_m at 16 °C.

12.2 Drug form, log *P* and Lipinski rules

The original published research considered by many to be fundamental to teaching in biopharmaceutics (Lipinski *et al.*, 1997; Lipinski, 2000) is discussed here and elsewhere (Suzuki *et al.*, 1998; Sarker, 2006a, 2012a; Cevc and Vierl, 2010; Benson and Watkinson, 2012). Lipinski's work can be used to evaluate proper drug working (movement across mucosae and epithelia; see Figure 12.6) or to determine if a moiety has a certain pharmacological or biological activity. The work is referred to as the 'rule of fives' (all numbers are multiples of five) and can be used to give an idea of *in vivo* Adsorption (Figure 12.6), Distribution, Metabolism and Excretion (ADME) of the drug (Figure 12.7).

Lipinski's rule states in general that, in order to be physiologically active, a drug must satisfy the majority of the following criteria:

- Less than 5 hydrogen bond donor atoms (nitrogen or oxygen with one or more hydrogens).

- Less than 10 hydrogen bond acceptor atoms (nitrogen or oxygen).

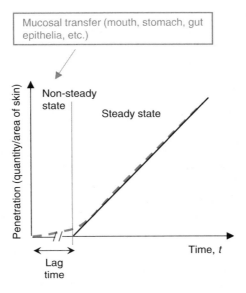

Figure 12.6 Mucosal drug delivery and uptake, involving a lag phase to a significantly high drug flux as a precursor to steady-state drug penetration

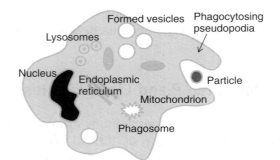

1. Pro-monocyte (bone marrow)		
2. Monocyte (blood) - nanoparticle removal		
Organ	3. Tissue macrophages (highly phagocytic)	Major site of organ nanoparticle removal
Liver	Kupffer cell	Yes
Lung	Alveolar macrophage	Yes
Spleen	Free and fixed macrophages	Yes
CNS tissue	Microglia	
Bone marrow	Sinus macrophages	Yes
Lymph node	Free and fixed macrophages	Yes

Figure 12.7   Combined schematic and table showing the organs and key macrophages involved in orchestration of nanoparticle clearance and destruction in the body's monocyte–phagocyte system (MPS; reticuloendothelial system, RES) defence mechanism

- A molecular mass (MW) less than 500 Da.

- A log *P* value of less than 5.

Other key physicochemical influences affecting cellular metabolism (see Figures 12.6, 12.7 and 12.8) not mentioned directly by Lipinski include moiety charge, molecule polar functionality and substitution and pK_a (Ghose *et al.*, 1999; Freitas and Müller, 1999). Molecules with a greater degree of unsaturation and aryl rings tend to have a higher lipophilicity (Table 12.2). These lipophiles in the circulation are then modified to more polar entities for excretion. Conventionally, *in vivo* this metabolism involves redox enzymes and the conversion of the drug to other, more hydrophilised, suitably modified compounds, and this process takes place mostly in the liver. However, some similar and supplementary processes also occur in the cells of the intestinal wall (see Section 12.4.1), the lungs and the blood compartment.

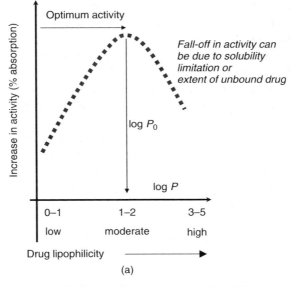

(a)

Lipinski's rule of 'fives'		
Structure	Molecular weight	Log *P*
Ionisation	p*K*a	
The molecule	Hydrogen bond donor	Hydrogen bond acceptor

(b)

Figure 12.8   (a) Optimum log *P* and drug polarity and the influence on *in vivo* absorption. (b) Lipinski's rules: tenets of bioavailability and drug uptake for a drug as one of the key parameters considered in product preformulation

One is able to clearly define an optimal log *P* (Figure 12.8) based on the local environment for delivery, the drug's structure (Figure 12.8b) and the perceived passage of the drug from its point of administration.

In order to reach a target tissue (Gregoriadis, 1977), a drug or drug nanoparticle absorbed into the bloodstream – via mucous (or endothelial) surfaces as found in the digestive tract (intestinal absorption) – is made therapeutically available after being taken up by the target cells. This can be a significant biopharmaceutics problem at some natural barriers, such as the blood–brain barrier (BBB; Hu and Dahl, 1999; Seelig, 2007; Maeda *et al.*, 2009). Factors such as poor compound solubility, delayed uptake (e.g. gastric emptying time, intestinal transit time; see Figure 12.6), chemical instability (e.g. in the stomach; Noguchi *et al.*, 1975; Brusewitz *et al.*, 2007) and an inability to permeate the intestinal wall can all reduce the extent to which a drug is absorbed after oral administration (Corveleyn and Remon, 1998; Hansen *et al.*, 2005). Absorption critically determines the drug compound's bioavailability. Drugs that absorb poorly when taken orally must be administered in some less desirable alternative manner. Once *in situ*, any drug excretion occurs at three sites within the body. The kidney is probably the most important, and is where products are excreted through the urine. Biliary or faecal excretion is the process initiated in the liver; it passes through to the gastrointestinal (GI) tract until the products are excreted and ejected. The third route of conventional excretion is through the lungs.

Higher lipophilicity (Table 12.2) favours particle association. Along with the risk of flocculation and an increase in particle size on storage or *in vivo*, this means that the physical form (in terms of the nanoparticle) must be considered carefully (Table 12.3). The ζ-potential for a drug delivery system (DDS) nanoparticle determines whether particles flocculate to produce an inferior product. Good shelf

Table 12.3 Commonly observed ζ-potentials for a range of particles

Sample	Value	Conditions
GOOD stability is observed	$>\pm 50\,mV$	$25\,°C$
Flocculation is usually seen	$<\pm 20-50\,mV$	$25\,°C$
Considered a 'highly charged' species	$>\pm 25\,mV$	$25\,°C$

Examples		
Non-ionic Pluronic F68 micelle	0 to $+1\,mV$	$25\,°C$, pH 7/$\mu = 0.05\,M$
PEGylated DMPE liposome	$+6\,mV$	$25\,°C$, pH 7
Sodium dodecyl sulphate micelle	-75 to $-90\,mV$	$25\,°C$, pH 7
Cetyltrimethyl ammonium bromide micelle	$+70\,mV$	$25\,°C$, pH 7/$\mu = 1.0\,M$
Atorvastatin microemulsion (glyceryl monostearate/ Tween 80)	$-9\,mV$	$25\,°C$, pH 7
Cationic liposome (DOPE)	$+40\,mV$	$25\,°C$, pH 7
Pickering-stabilised emulsion (alumina)	$+60\,mV$	$25\,°C$, pH 7
Pickering-stabilised emulsion (silica)	$-40\,mV$	$25\,°C$, pH 7
Intravenous fat emulsion	$-50\,mV$	$25\,°C$, pH 7
Lecithin-coated emulsion	$+40\,mV$	$25\,°C$, pH 7/$\mu = 0.01\,M$
Serum albumin emulsion	$+30\,mV$	$25\,°C$, pH 3/$\mu = 0.05\,M$
β-lactoglobulin-coated food emulsion	$-40\,mV$	$25\,°C$, pH 7/$\mu = 0.01\,M$
Sulphonated polystyrene latex	$-30\,mV$	$25\,°C$, pH 7/$\mu = 0.05\,M$

μ, ionic strength, phospholipids; DMPE, dimyristoyl phosphatidylethanolamine; DOPE, dioleoyl phosphatidylethanolamine.

life and therapeutic 'stability' (see Section 3.2) is conferred on the particle by either large negative or large positive values, thus facilitating electrostatic repulsion between particles. For optimal therapeutic performance, emulsifier selection and measurement of ζ-potential are important preformulation considerations.

12.3 Skin and epithelial models

The skin is sometimes referred to as the largest organ in the human body (area up to $\sim 2.0\,m^2$). The 'organ' is composed of the upper stratum corneum (SC), middle dermis, subcutaneous fat and lower basal skeletal muscle. The lower portions are extremely well supplied with blood and are highly vascular. Drugs delivered topically to the SC or injected intradermally (ID), subcutaneously or intravenously (IV), or as a depot emulsion, find their way, after a lag into the blood compartment (Figure 12.6), to their desired location. The skin and the cells therein are approximately 80% water, and this means that oily vehicles form an occlusive film and do not pass easily into the skin. The remaining 20% is made up of 70% fat and 30% protein (keratin and collagen). The composition is highly variable with the depth of positioning in the skin. Dermatological scientists (Moghimi, 1996) have proposed a compositional brick wall model (see Figure 8.1), which explains the skin heterogeneity. In some cases, where drug passage is poor, drug formulators use penetration enhancers (cosolvents) such as dimethyl sulphoxide (DMSO) or propylene glycol (Busse, 1978), polar oils or ethanol, which facilitate drug passage deeper into the skin.

The epithelia found throughout the body (see Figure 7.3b) are a complex network of interconnected excretory basal cells linked by a paracellular gap or tight junctions (Hu and Dahl, 1999; Seelig, 2007; Lemke *et al.*, 2013) and an extracellular matrix, which generate an outer wettable surface of mucilage (hydrophilic glycoprotein and exogenous excretions). The epithelium serves as a protective surface in an internalised cavity and has a lubricatory role within the GI tract and vagina, but it may also be used to facilitate drug passage into the blood and neighbouring tissues (see Figure 7.3). While the skin has a large surface area (Souto *et al.*, 2004), it is noteworthy that the epithelial surface areas of $100-160\,m^2$ for the lung and $300-400\,m^2$ of the colon are hundreds of times greater. This indicates the reasoning behind current targeting, and a potential margin and scope for the therapeutic use of these organs and their surfaces.

Macromolecular drug and nanoparticle vehicle (emulsion, liposome, micelle, nanoemulsion, SLN) passage across and within both the SC and the epithelia is curtailed by size constraints. These contraints include the endo- and transcytosis vesicles of the cells, their receptor-driven initiation and the tight junctions between neighbouring cells (see Figure 7.2). This respective transcellular and paracellular passage of drugs has a significant role to play in drug delivery science and optimal efficacy through careful drug design (Müller *et al.*, 2000).

12.4 Drug delivery routes

Where drugs are encapsulated in an emulsion or lipidic particle, their release from the oil phase is one of the barriers (or opportunities) to product control and efficacy. The primary routes for dispersions and nanoparticles are the ocular, otologic, nasal, pulmonary, sublingual, buccal, oral–gastric–enteric (jejunal, duodenal, ileal), subcutaneous implant, vaginal, colonic or rectal, topical, transdermal and parenteral. Parenteral routes include IV, subcutaneous, intramuscular (IM), intraspinal (ISp), intraosseous (IO), intrasynovial (ISy) and intrathecal (IT) delivery. Most high-potency drugs (e.g. cytotoxic drugs) are delivered by the parenteral route, customarily by IV line or IV injection, and orally, but this is not feasible for emulsion-type products without frequently breaching particle integrity (see Chapter 3) and compromising the dosage form consistency and thus the efficacy. Across all types of pharmaceutical products, oral delivery accounts for ~50% of dosage forms, parenteral ~25%, topical ~10% and others (nasal, pulmonary, rectal) ~15%. This picture changes significantly for emulsion and colloidal particle systems, where oral delivery accounts for merely ~5% of dosage forms, parenteral a hugely significant ~55%, topical ~30% and others (nasal, pulmonary, rectal) ~10% (Müller *et al.*, 2000).

12.4.1 Mucosal delivery

Targeting of any tissue usually requires that a drug be taken into the bloodstream conventionally, via mucous surfaces (see Figure 7.3b) such as the nasal passage (Kaur *et al.*, 2008; Kumar *et al.*, 2008), oral space, buccal cavity, urethra, vagina (Henzl, 2005; Jones and Harmanli, 2010) or GI tract, via ileal, colonic or rectal (Nishioka *et al.*, 1980; Suzuki *et al.*, 1998) absorption (Cavalli *et al.*, 2002). The therapeutic effect of the drug is then seen after it is taken up by the target cells. This easy uptake is problematic with some organs with significant natural physiological 'barricades' (Figure 12.7), such as the BBB, where small pore sizes (3–15 nm, 500 Da) and intrinsic protective mechanisms particular to this structure do not favour para- or transcellular passage (Hu and Dahl, 1999; Maeda *et al.*, 2009) or drug movement (see Section 12.4.3). Passage across mucosae is initially limited by a lag before a steady-state flux of drug (see Figure 12.6) comes into effect.

12.4.2 Systemic

This type of DDS involves injection or passage directly into systemic circulation and portal passage via a Hinkman line. Any particles or oil droplets injected must be <5 μm in diameter, since an erythrocyte is approximately 6 μm and the diameter of a fine capillary is approximately 20 μm. Nowadays, most productions aim for <1 μm average droplet size, and most nanoparticle DDSs for 0.1–0.2 μm. Smaller droplets usually mean less chance of vascular occlusion, clot formation, emboli and rapid macrophage clearance (see Figure 7.3a and Figure 12.7) from the blood compartment (Davis *et al.*, 2008; Araujo *et al.*, 2011).

12.4.3 Oral (gastrointestinal)

After swallowing the vehicle, the drug passes to the stomach. The linings of the throat, oesophagus and stomach are mucosal in nature. The drug is absorbed through these surfaces but the bulk of the absorption occurs when the drug traverses the gastric or ileal mucosae (GI tract). The dissolution apparatus (US, British, European and Japanese Pharmacopoeia; USP, BP, EP, JP) has long been used to model disintegration and solubilisation in the gut. The method is used routinely for solid dosage forms (tablets, capsules), but also for suppositories and fat-based pessaries. Similarly, SC or epithelial absorption and drug efflux (see Figure 12.6) can be modelled using cadaver/animal skin or an artificial membrane isolated from the two chambers of the Franz diffusion cell. This approach is used widely for topical and colonic DDSs. To date, limited use of the oral–GI tract route has been underpinned by concerns over product consistency (see Chapter 3). Emulsified products simply do not resist the harsh environment of the stomach, which causes emulsifier surface instability and coalescence/flocculation. Nutraceuticals and vitaminised products or silicone oil antireflux formulations do, however, use this vehicular form quite frequently. A new wave of novel controlled, sustained and modified release gel tablets (Corveleyn and Remon, 1998; Hansen, *et al.*, 2005) with direct use of SLN-encapsulated (Müller *et al.*, 2000; Pouton, 2000; Moses *et al.*, 2003; Chen *et al.*, 2006; Sarker, 2012a) 'emulsified' drugs may open the gastric route to more forms of product (Constanides, 1995; Brusewitz *et al.*, 2007).

When drug targeting via the oral route is used (see Table 12.4), factors such as high log *P*, poor gastric emptying (or local entrapment), small

Table 12.4 Well-known medical products based on coarse emulsions for oral administration

Emulsion	Drug	Commercial name	Therapy
Castor oil	–	–	Constipation
Corn oil, castor oil	Penicillin V,G	Sandimmune/	Antimicrobial
Olive oil	Amoxicillin	Neoral	Immuno-
	Streptomycin		suppression
	Cyclosporin-A		
Castor oil	Ritonovir	Norvir Fortovase	HIV therapy
glycerides	Saquinavir		

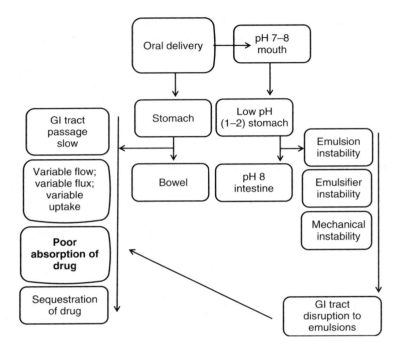

Figure 12.9 Oral delivery and its use of emulsions. Major issues limiting use of this route revolve around harsh pH, high ionic strength, complex interactions and mechanical perturbation of the sample and its effect on the nature of the dispersed particle

and large bowel transit times, GI-induced chemical instability (e.g. lipolysis/hydrolysis/esterification) or peptide/polysaccharide binding and catalysis (and ionisation) and an inability to permeate the intestinal wall can all reduce the extent to which a drug is absorbed following oral administration (Figure 12.9). Absorption critically determines a compound's bioavailability; consequently, if this is in doubt, other more reliable routes are generally considered as these confer a greater degree of reliability and PCQ (purity, consistency, quality).

Emulsions find use in a range of medicinals and medical materials. They are fabricated for internal and external use, primarily carrying apolar, noxious, toxic or labile drugs.

13

Sizing and microscopy

Product performance and suitability is tied to the size, morphology (symmetry, roughness, surface coverage) and topology of dispersed particles (Figure 13.1). As described earlier, particle irregularity can promote coalescence, Ostwald ripening, creaming, flocculation and sedimentation, and also alter the product rheological characteristics (Sarker, 2005a). Viscoelasticity and viscosity change, for example, alter topical application (i.e. spreadability) and thus the administered dose, but may also alter dissolution characteristics. An essential part of ensuring quality (Sarker, 2012a), efficacy and consistency lies in a fuller morphological characterisation. Microscopic assessment, such as transmission electron microscopy (TEM), scanning electron microscopy (SEM), atomic force microscopy (AFM) and the hot-stage polarising microscopy (HSPM) are useful for providing visual imagery of particles (birefringent polymorphs appear as differing colours with HSPM) but do not describe charge or the water sheath carried by the particle (Al-Hanbali *et al.*, 2006; Georgiev *et al.*, 2007; Concannon *et al.*, 2010). Some exceptionally specialised forms of microscopy (interferometry, fluorescence recovery after photobleaching (FRAP)) are described in detail in Sarker *et al.* (1995a, 1996) and Sarker and Wilde (1999).

13.1 ζ-potential

The surface charge of a particle with an associated counterion (i.e. shear plane) sheath is referred to as the ζ (zeta)-potential (mV). The charge at the end of the particle is known as the Nernst potential. ζ-potentials (see Table 12.3) are deduced by Doppler anenometry (Al-Hanbali *et al.*, 2006) using the Smoulochowski equation. The technique works by measuring both the particle surface voltage in response to a complimentary and opposing equivalent applied external field and the cell obscuration that this causes to the 632.8 nm HeNe laser beam when particles 'crowd' in a particular spatial position in the photocell at 25 °C.

Pharmaceutical Emulsions: A Drug Developer's Toolbag, First Edition. Dipak K. Sarker.
© 2013 John Wiley & Sons, Ltd. Published 2013 by John Wiley & Sons, Ltd.

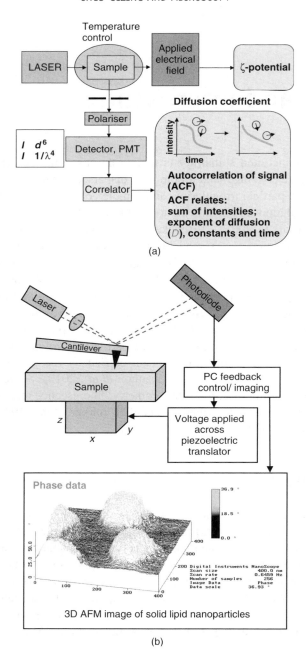

(a)

(b)

Figure 13.1 Methods of particle size and form evaluation. The figure is complex and presents two different methodologies that are often used simultaneously. (a) Dynamic light scattering (photon correlation spectroscopy, PCS), which is often additionally linked to ζ-potential measurements of surface charge. (b) Basic equipment and a specimen of a 3D image captured from an atomic force microscope (AFM) of solid-state spheroidal lipid particles. AFM phase data can be used to infer nanorheological textural information

13.2 Hydrodynamic diameter

The particle size, with its permanent sheath of bound water, is estimated from the Stokes–Einstein equation according to well-utilised standard methodologies such as Mastersizer/Nanosiser/Zetasizer or Contin (Sarker *et al.*, 1995a) programmes for low-Reynolds-number fluids (low viscosity, $\eta \sim 1.0$ mPas) with spheroidal particles (Al-Hanbali *et al.*, 2006) of radius r_A:

$$D(m^2/s) = k_B T/6\pi \eta r_A \tag{13.1}$$

$$\text{radius}(r_A) = k_B T/6\pi \eta D \tag{13.2}$$

Maxwell and Mie theory mathematical solutions are used to find the bulk diffusion coefficient (D) from an expression with the correlation coefficients and cumulants (e.g. G_1 and G_2) used to describe the time-averaged autocorrelation of the dynamic pattern of light scattering at 90° with respect to the incident laser light beam (632.8 nm HeNe) at 25°C. Size is related to the diffusion coefficient by Equation 13.2. Large particles (MW $= 20$ MDa) such as viruses (influenza) have a D of 0.3×10^{-11} m^2/s, whereas small particles such as oligopeptides (MW $= 300$ Da) have a D of 130×10^{-11} m^2/s. Two indices are very useful for approximately spheroidal oil (solid, self-assemblies) or water droplets: the $d_{3,2}$ (surface area-averaged diameter; Sauter mean) and the $d_{4,3}$ (volume-averaged diameter), which are used in the main even though other, less useful forms do exist.

14
Rheology, texture, consistency and spreadability

Predictable rheological performance (the science of material flow) is both crucial and imperative in any high-quality pharmaceutical product. The product flow properties become pivotal to use and therapeutic treatment in just the same manner as drug content and purity are key considerations.

Texture evaluation in terms of Young's modulus of elasticity and numerous forms of viscosity (and viscoelasticity), determined for example by the couette or cone-and-plate viscometer (Figure 14.1), are used and an indicator of physical consistency (e.g. hand-feel, mouth-feel, tackiness).

14.1 Bulk properties

The coefficient of viscosity (η), hereafter referred to as the viscosity (units: $1\,\text{mPas} = 1$ centistoke (cSt) $= 1$ centipoise (cP)) is a measure of the internal friction within a sample and its resistance to shear (γ_{dot}). Lower viscosities are equal to low internal friction within the sample. In an emulsified product, increasing the number of dispersed oil droplets therefore increases the viscosity. The viscosity may also be defined as the shear stress (τ; Pa) acting on a system defined according to Newton's model:

$$\tau = F/A = (\eta \times \text{velocity gradient}) \tag{14.1}$$

where F is force, A is area and the velocity gradient or rate of shear is du/dy. Here, du is the planar velocity and dy is the distance in the sample over which this velocity is seen. The simplest way to define dynamic viscosity is as follows:

$$\eta = \tau/\gamma_{\text{dot}} \tag{14.2}$$

Pharmaceutical Emulsions: A Drug Developer's Toolbag, First Edition. Dipak K. Sarker.
© 2013 John Wiley & Sons, Ltd. Published 2013 by John Wiley & Sons, Ltd.

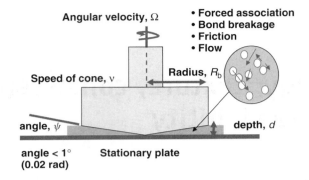

Figure 14.1   Cone-and-plate rheometer for constant shear dynamic viscosity and oscillatory rheometry for storage (elastic element) and loss (viscous element) moduli. The cone-and-plate rheometer is the mainstay of texture analysis for semisolids and gels

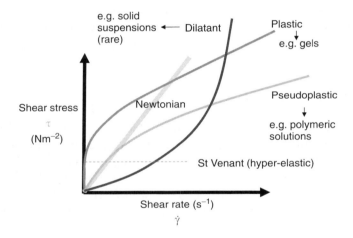

Figure 14.2   Bulk rheology and the basic forms of viscosity seen with fine and coarse dispersions. The basic types are Newtonian, plastic (Bingham), pseudoplastic, dilatant and hyper-elastic (St Venant). The sample viscosity profile can be used for quality control (QC) and to define or predict the compositional elements of the product

There are four basic forms of viscosity (shown in Figure 14.2); these time-independent processes (Jones, 2002; Sinko, 2006) are:

- *Newtonian system*   Shear thinning, the simplest form of flow; only found in 'pure' solutions (e.g. liquid paraffin) or very dilute (low phase volume) dispersions of small, regularly shaped particles (e.g. micellar suspension at critical micelle concentration, CMC).

- *Bingham or plastic system*   Shear-induced thinning, with a characteristic yield stress (τ_B) at which 'breakage' or 'slippage' occurs (Bingham stress). Typified by concentrated, high-phase-volume ($\phi > 0.5$) pastes, slurries and dispersions. Very common across an enormous range of pharmaceutical materials, e.g. oil-in-water (O/W) or water-in-oil (W/O) creams (e.g. acyclovir cream), fat-based suppositories, polymer composites and semisolids.

- *Pseudoplastic system*   Shear-induced thinning. System constituted by temporary and weaker intermolecular associations. Typified by polymer gels in which the nature of crosslinking is generally not covalent in nature and/or the dispersed phase polymer/droplet ($\phi \approx 0.1$) is moderately high, and therefore disrupted by low shear rates. Partial disruption occurs immediately upon agitation. e.g. Bonjela, a hypromellose-lidocaine gel.

- *Dilatant systems*   Shear-induced structure forming (independent of time). Usually implies a sample constituted from irregular particles. Not seen routinely but examples include starch granule slurries and slurries of silicates.

Time-dependent rheological (Sinko, 2006) behaviours involve complex interrelationships. Examples are:

- *Thixotropic*   These characteristics are shown by some samples, such as pharmaceutical creams. These are time-dependant shear-induced thinning materials that get progressively less viscous as the duration of deformation proceeds. The consistency is 'weakened' or 'loosened' following agitation or spreading.

- *Rheopectic*   Time-dependent (and shear-induced) change in viscosity. These thicken (e.g. crosslinked hyaluronic acid) or solidify when shaken (the opposite of thixotropy).

Other key rheological forms include:

- *Elasticity*   Hookean structures act like 'springs' and return to their original form after a deforming stress has been removed (e.g. crosslinked gums or crosslinked nanogels, or St Venant, i.e. hyper-elasticity, where deformation and stress is localised and can be compensated internally).

- *Viscoelasticity*   Voigt and Maxwell models (see Figure 14.3), dashpot and spring elements. These materials show a mixture of viscous (energy dissipative, G'') and elastic (energy storage, G') characteristics in various combinations (e.g. aggregates and extended-polymer-chain combinations adsorbed on the surfaces of oil droplets).

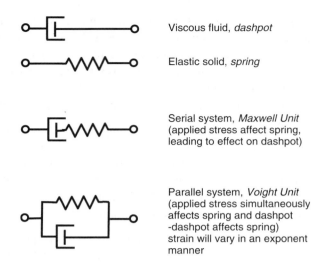

Figure 14.3   In addition to the viscous nature of dispersions, many composite materials that are mixtures of regular particles and extended polymer molecules show viscoelastic behaviour. The various elements of viscoelasticity are presented in the figure. Most viscoelastic (interconnected complex network) samples can be considered to be singular or multiple compositions of the fundamental units shown here

14.2 Solid-state and nanorheological properties

The harder semisolids (e.g. solid lipid nanoparticles (SLNs) or proper solids) and composite materials can be investigated by solid-state techniques such as atomic force microscopy (AFM). These techniques permit the mechanical and nanorheological properties (Al-Hanbali *et al.*, 2006) of the material to be examined at resolutions of tens of nanometres. Techniques such as 'tapping-mode' AFM can be used to measure the roughness or topology and phase lag. Phase lag can be used to give an indication of hardness, elasticity and softness (Concannon *et al.*, 2010) and has been used to demonstrate polymer brush or mushroom surface configuration (Al-Hanbali *et al.*, 2006). Other texture-analysis apparatuses, such as Microsystems TA-TXT (or thermal analysis techniques), can be used for direct measurements of hardness/compressibility, defects, tensile strength, tackiness, extensibility/elongation, break point/break or yield stress, brittleness and Young's modulus of rigidity (slope of stress/strain). These are illustrated in Figure 14.4.

14.3 Interfacial properties

Examination of the properties of polymers, drugs and emulsifiers located on the surfaces of solids and oil or water droplets can be modelled by reference to 'equivalent' macroscopic interfaces (Figure 14.5). The vast majority are usually

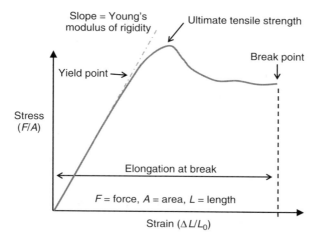

Figure 14.4 A stress–strain plot demonstrating mechanical texture, its evaluation and key composite material parameters such as Young's modulus of rigidity, ultimate tensile strength and the break point of a solid or semisolid material

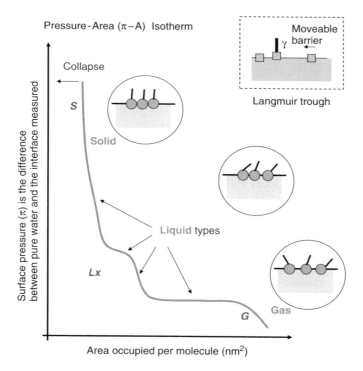

Figure 14.5 Surface structure and spacing in surface-adsorbed emulsifier layers, determined by the Langmuir trough (LT) and used to fabricate Langmuir–Blodgett (LB) films. Materials constrained at the interface show 2D behaviour that is similar (but not identical) to 3D macroscopic behaviour

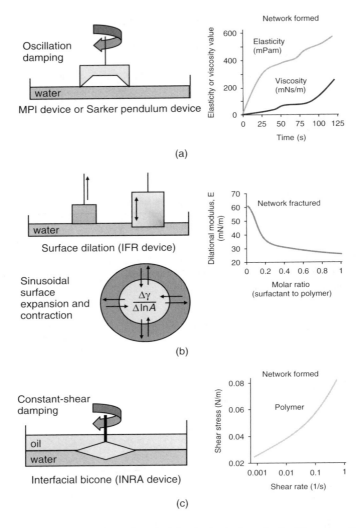

Figure 14.6 A trinity of surface (interfacial) rheological methods. There are three distinct methods: (a) an A/W shear rheological device, originally conceived at the Max Planck Institute, Berlin, Germany; (b) an A/W surface-dilational (dilatational) apparatus, conceived at the Wageningen Agricultural University, Wageningen, Netherlands/Institute of Food Research (IFR), Norwich, UK; and (c) a constant-shear bicone device used for probing O/W interfaces, devised at the Institute Nationale de la Recherche Agronimique (INRA), Nantes, France. The interfacial structure can be used to predict and 'engineer' better drug particle integrity (see Figure 14.7)

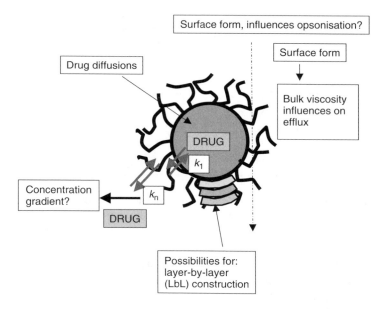

Figure 14.7 Drug particle encapsulation technology. This may include charged or noncharged surface active polymers and low-molecular-weight (LMW) surfactants, or mixtures. Newer 'platform' approaches (other than by spontaneous self-assembly) include the use of layer-by-layer or layer-on-layer (LbL) surface construction. This can be used to modify the drug release kinetics and mechanical or stability performance of the product (see Sarker, 2010)

undertaken on water in contact with air or nitrogen (Sarker *et al.*, 1995a,b, 1999); occasionally this is extended to look at the properties of polymers at the oil–water interface (Clark *et al.*, 1991a,b; Wilde and Clark, 1993). Due to their low dielectric constants (apolarity), but disregarding dissimilar viscosity–densities, air and oil can be viewed as sharing some essential characteristics and so may be studied in parallel (Figure 14.6). Self-assembled aqueous structures such as thin liquid films (TLFs; Sarker *et al.*, 1996; Sarker and Wilde, 1999; Sarker, 2005b, 2009a,b) can be used to gain an insight into surface emulsifier structure (or structure in the liposome leaflet), shown in Figure 14.7.

The principal methods for evaluation of interfacial properties are indicted in Figure 14.6. The most useful are surface rheology (dilation (dilatation), shear, pendulum apparatus), surface tension (Figures 2.8 and 14.5), fluorescence recovery after photobleaching (FRAP) and other TLF-based microscopy (Figures 3.4 and 3.5) imaging methods (Sarker *et al.*, 1996, 1998b).

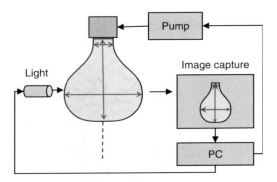

Figure 14.8 Drop shape and image analysis for the measurement of surface tension and surface mechanical behaviour. Newer equipment can predict surface tension to a resolution of ~0.01 mN/m and provide information on the storage (G') and loss (G'') moduli

Today, image analysis of pendant and sessile droplets (Figure 14.8) is also being used routinely for mechanical evaluation of interfacial elasticity and viscosity (and interfacial tension). The advantage is that it can be undertaken easily on fluid interfaces. The speed of analysis and the low volume of samples needed are also favourable for research purposes or where little sample is available and the technique fits better with routine quality control (QC) due to its physical robustness.

15

Quality control, process analytical technology and accelerated testing

The primary advantage of process analytical technology (PAT) is its enabling of real-time release (RTR) *in situ* (Howbrook *et al.*, 2002, 2003; Sarker, 2004a,b; van der Valk *et al.*, 2003). The standard forms of PAT are represented in Figure 15.1. Examples include:

- Metrohm Rancimat 743 equipment (Metrohm Instruments), which performs oxidative stability tests based on conductivity and is used on oils, fats and 'lipidics'.

- Metrohm 763 PVC thermostat, a materials degradation tester.

- Colourimetry, e.g. thiobarbituric acid for the products of lipid autoxidation, such as malondialdehyde.

PAT generally means a reduction in sampling time (Capelle *et al.*, 2007) and an improvement in efficiency, which facilitates process continuity (Sarker, 2004a,b). Accelerated testing is really about exposing the product to more extremes (of the conditions that instigate degeneration) in a short period, 'equivalent' to many years' worth of exposure. Emulsions and lipidic samples such as vesicles are subject to autoxidation as a result of ultraviolet (UV) light/lamp exposure, oxygen and heat, and this can have dire effects on the chemical integrity of the product and the rate of degradation (Sarker, 2008).

Accelerated testing is used to screen components of the formulation and of production equipment employed during the manufacture of pharmaceutical emulsions and related products. The benefits of accelerated testing regimes include the ability to estimate the lifetime (shelf life or viable shelf life) of the product and enabling the manufacturer to find, improve and control the active pharmaceutical ingredient (API), components, materials and processes key to successful product PCQ (purity, consistency, quality).

Pharmaceutical Emulsions: A Drug Developer's Toolbag, First Edition. Dipak K. Sarker.
© 2013 John Wiley & Sons, Ltd. Published 2013 by John Wiley & Sons, Ltd.

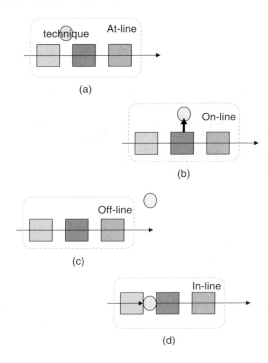

Figure 15.1 Different forms and levels of sophistication in process analytical technology (PAT) used to follow and aid quality maintenance in industrial products. Examples of associated analytical technologies include built-in spectrophotometers and photoacoustic sensors. There are four scenarios: (a) at-line, (b) on-line, (c) off-line and (d) in-line

15.1 Preformulation, high-throughput screening

Preformulation is part of the drug development process (see Table 12.1). It involves matching knowledge of basic chemistry (Howbrook *et al.*, 2002, 2003; Sarker, 2004a,b; van der Valk *et al.*, 2003) and of physical chemistry (pK_a, log P, hydrolysis, polymorphism, T_m, solubility, photosensitivity, oxidation) to potential behaviour *in vivo* (Sarker, 2012a). Preformulation features significantly at two key points: the first is at phase 0 and preclinical testing and the second is at the point of phase III–IV, where the design of the drug for administration and the industrial scale-up of both the manufacturing process and current good manufacturing practice (cGMP) come into effect. As in all GMP, validation and risk analysis of both materials and assembly to a consistent and predictable finished product are essential if the medicine is to match the product licence (see Figure I.1). Validation and 'mapping' of uncertainties is the point of weakness for most manufacturing. This is discussed at length and in all its common forms in Sarker (2008). Validation and high-throughput screening

(HTS) involve fuller characterisation of the desirable chemical traits (Howbrook *et al.*, 2002, 2003; van der Valk *et al.*, 2003) of the excipients, drug actives and combinations. Again, basic chemical properties are used to determine the optimal form of therapeutic delivery. Solubility, chemical stability and biopharmaceutic performance, which determine parameters such as pK_a and log P (Lipinski *et al.*, 1997), require clarification in order to ideally formulate and deliver the drug within a biocompatible vehicle. For drug products based on macromolecules and simple organic drugs, key considerations are addressed well by HTS.

Preformulation (Sarker, 2004b) is a most important part of drug development (see Table 12.1). Validation of standardised manufacture or performance of the materials is the point of uncertainty and risk for most commercial products. The chemical traits that need elucidating for therapeutics are:

- biological and pharmacologic activity;
- chemical degradation and purity;
- physical form.

Standard methods for undertaking HTS include UV spectroscopy and the characteristic energy (wavelength, λ; frequency, ν) of the light waves absorbed, differential scanning calorimetry (DSC), circular dichroism (CD), infrared spectroscopy (FTIR) absorptions, many types of chromatography, Raman spectroscopy, fluorescence spectroscopy, mass spectrometry and ion-specific electrochemical techniques (Figure 15.2a). Candidate drug or product selection can make use of HTS in terms of excipient selection (Figure 15.2b), excipient compatibility and optimal solubility across a range of media, in terms of preformulation and the creation of a prototype of the form of the drug product. Key criteria would be selective release of the drug, *in vivo* optimisation, characterisation, viability and toxicology (chronic, acute and cumulative).

15.2 Industrial concerns

One of the main dilemmas in the consistent and safe mass production of heterogeneous systems such as emulsions, microemulsions, SLNs and liposomal products is how to ensure uniformity without compromising product performance or stability. Paul Ehrlich, who discussed therapeutic delivery in the 'magic bullet' form says, 'The first rule of intelligent tinkering is to save all the parts' (*Saturday Review*, 5 June 1971). This is difficult for heat-sensitive materials when some degree of intervention is required. Additionally, the consistency (Sarker, 2004b) of the polymorphic form of actives and excipients and the purity (safety) of raw materials also have a bearing on the assurance of quality and its routine control (Benoliel, 1999; Sarker, 2008).

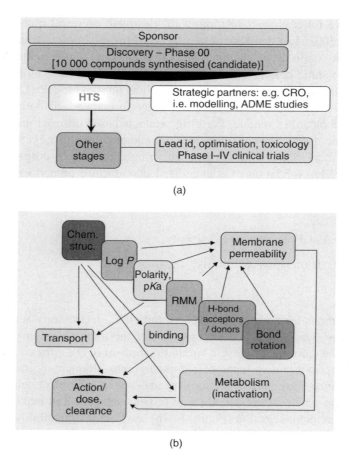

(a)

(b)

Figure 15.2 High-throughput screening (HTS) and its central role in the physicochemical characterisation of pharmaceutical actives and products in terms of drug optimisation, bioavailability, innovation and cost saving. (a) HTS's uses as part of drug product development (b) Key tests and parameters that HTS must account for in routine preformulation and continuing development

15.3 Rancimat and other methods

Standard methods such as the Rancimat are used to test the shelf lives of fat- and oil-based products (see Chapters 5, 13 and 14 and Sections 1.3, 2.1.2 and 6.2). This is important as both oxygen and light (and prooxidants such as trace amounts of heavy metals) have a large impact of the chemical stability of this type of product. In exposing a product to extremes of environmental 'abuse', scientists can more accurately predict product suitability and efficacy (Capelle *et al.*, 2007; Sarker, 2008). Most manufacturers also devise in-house methodologies to characterise their emulsion and colloidal products. Chemical stability impinges on physical stability (Russell, 1991; Ansel *et al.*, 1999; Lupo, 2001; Di Mattia *et al.*, 2010; Alayoubi *et al.*, 2013) and this modifies the suitability of the drug product. Key tests (Benoliel, 1999) will include a version of the accelerated testing of autooxidation and rancidity, active modification (e.g. protein denaturation, drug isomerism), particle flocculation and polymorph crystal form.

Questions

The author provides here a series of questions of varying difficulty, sophistication and complexity, designed to assist in learning and test broader understanding (Sarker, 2012b). Practice consists of practical and pragmatic real examples, conventional considerations, knowledge expected on reading around the subject, questions based on reasoning an intuition and a sampling of useful calculations, such as those of hydrophile – lipophile balance (HLB) and pK_a.

Guide for readers

The long questions should and do have scope for variation in the response. Standard expected elements of a high-scoring answer are indicated. When attempting the longer questions, you should consider what else might be written, and whether it captures state-of-the-art knowledge at this point in time.

Specimen 'test' questions

These are provided as an aid to the student, industrialist, researcher or newcomer to the area of industrial pharmaceutical emulsions.

Short

Short questions, requiring detail as indicated by the mark. Three marks roughly equates to three to four sentences.

1. What causes the surface tension (γ) of a pure aqueous liquid to arise? (3 marks)

2. The spreading of a liquid drop on a solid surface in air is a function of contact angle (θ) and what else? (3 marks)

Pharmaceutical Emulsions: A Drug Developer's Toolbag, First Edition. Dipak K. Sarker.
© 2013 John Wiley & Sons, Ltd. Published 2013 by John Wiley & Sons, Ltd.

3. Discuss the work of Alec D. Bangham, the invention of the liposome in Cambridge, and its use in modern medicine, from current use to state-of-the-art technology. (5 marks)

4. Critically discuss the invention of the niosome by l'Oréal in 1975 and its extension to modern pharmaceutics and biopharmaceutics. (5 marks)

5. Outline and both critically discuss and describe the technology of production and science of application of the solid lipid nanoparticle (SLN) and lipid nanocapsule (LNC) to drug delivery science. (5 marks)

6. Critically discuss the use of 'soft-matter science' in the fabrication and use of medical nanoparticles for the delivery of potent drugs. (5 marks)

7. Based on molecular structure, why are Spans (lipid sorbitan esters) rather than polysorbates (lipid polyoxyethylene esters) typically used for niosomes? (5 marks)

8. Briefly describe one example of each of the following: (i) micellar, (ii) polymer nanoparticle and (iii) lipid nanoparticle drug delivery system (DDS). (5 marks)

9. Discuss the three primary mechanisms behind gelation of block copolymer colloids. (5 marks)

10. What are the *three* fundamental factors that influence thin liquid film (TLF) and thus droplet metastability? (5 marks)

11. Briefly describe *three* of the ten or so drug delivery systems that are influenced or driven by surfactant behaviour and adsorption processes. (5 marks)

12. Discuss the contributions of the DLVO theory to particle stability. (10 marks)

13. Discuss the types, uses and principles of formation of an emulsion droplet and factors influencing product 'quality'. (15 marks)

Long

A selection of real, hypothetical and problematic scenario assessment materials for the reader, which should aid understanding of the field and integration of subject themes and provide an addition to material contained within the chapters. These questions are to be expanded upon based on an approximate marking scheme (provided later). The information needed to answer these questions is contained in relevant subsections within the book.

1. In terms of pharmaceutical analysis, critically discuss the essential uses of thermal analysis by differential scanning calorimetry (DSC) and the technique's primary applications, strengths and weaknesses. (30 marks)

2. Discuss the use and effect of emulsion formulation aids in the manufacture of PCQ (purity, consistency, quality) products. (50 marks)

3. You are provided with two alternative procedures for improving the quality of a topical oil-in-water (O/W) product:

   • Improve the manufacturing procedure according to ISO 9000/2001.

   • Reduce, using appropriate procedures, the number of defective and poorer-quality (nonconforming) raw materials used in an unchanged manufacturing procedure.

   Discuss, with reasons, which of these two approaches should be adopted in order to improve the quality of the final product. (50 marks)

4. Discuss the basis of the 6-sigma control and quality system and relate this to routine process control for the production of a sterile colloidal DDS. (50 marks)

5. A pharmaceutical manufacturer is routinely producing a sterile biopharmaceutical product. The nature of the product is an aqueous, buffered, ampoule-based form of an encapsulated protein drug. Describe the concerns surrounding routine effective good manufacturing practice (GMP)-grade production and the remedial action, if any, a manufacturer of bulk biopharmaceuticals might take to ensure the best-quality product. (50 marks)

6. Discuss the concepts behind quality assurance of parenteral medicines. Include in your answer examples of its relevance to patient care in a hospital. (20 marks)

7. What pharmaceutical/chemical data are required to provide a good understanding of the characteristics of a new drug substance prior to formulation into a product? Show how these data can provide guidance in minimising formulation difficulties. (50 marks)

8. Put yourself in the position of the developer of a new drug/medicine or dosage form. What are the key steps and critical criteria involved from concept to routine prescribed product? (50 marks)

MCQs

1. What is a Pickering emulsion?

   A. An emulsion stabilised by ice.

   B. An emulsion stabilised by solvent.

   C. An emulsion stabilised by solid particles.

   D. An emulsion stabilised by surfactants.

   E. An emulsion stabilised by drugs.

2. What is competitive adsorption or competitive displacement, with reference to the surface adsorbed layer?

   A. Displacement of a polymer by a small amphiphile.

   B. Displacement of a small amphiphile by a polymer.

   C. Displacement of an acid by a base.

   D. Displacement of a hydrophobe by a hydrophile.

   E. Displacement of a hydrophile polymer by a hydrophobe polymer.

3. Which of the following is true upon increasing surfactant concentration?

   A. Micellisation is less likely.

   B. Surface viscosity is unaffected.

   C. Surface viscosity goes down.

   D. Surface tension goes up.

   E. Surface tension goes down.

4. Based on the name of the Cambridge University scientist who discovered the liposome in 1961 – 1964, which of the following is a euphemism for a vesicular system?

   A. Singersome.

   B. Bangasome.

   C. Bestosome.

   D. Erlichasome.

   E. Sangersome.

5. What is the phase transition temperature for the liposome phospholipid, DMPC, likely to be if DLPC, DPPC and DSPC are 0, 43 and 57 °C, respectively?

    A. −12 °C.

    B. 24 °C.

    C. 67 °C.

    D. 48 °C.

    E. 43 °C.

6. What is the usual ratio of nonionic surfactant to cholesterol to adjunct surfactant in a niosome required to provide structural integrity and appropriate intercalation of bilayer 'stiffeners' and 'functionalisers'?

    A. 0.5 : 1 : 1.

    B. 3 : 3 : 2.

    C. 1 : 1 : 3.

    D. 3 : 4 : 5.

    E. 3 : 1 : 1.

7. Fusogenic vesicles (niosomes or liposomes) work by changing from bilayer to _____ phase, thus obviating temporary membrane dissolution.

    A. Hexagonal, via surface charges and lyomorphic change.

    B. Tetragonal, via surface potential and thermomorphic change.

    C. Laminar, via surface charges and lyomorphic change.

    D. Trilayer, via surface charges and mesomorphic change.

    E. Hexagonal, via surface charges and thermomorphic change.

8. Niosomes have better skin penetration than liposomes because of which of the following?

    A. Apolarity and dehydration.

    B. Shape and hydration.

    C. Size, polarity and hydration.

    D. Larger size.

    E. Charge and ζ-potential.

9. Who invented the niosome in a 'ground-breaking' patent in 1975?

    A. University of Santa Barbara.

    B. MIT – Harvard University.

    C. Unilever.

    D. BASF.

    E. L'Oréal.

10. The solid lipid nanoparticle (SLN) was first invented by (i) whom and was mass manufactured and purified by (ii) whom, respectively in two patents?

    A. (i) Müller and Luck, 1993; (ii) Gasco, 1993.

    B. (i) Miller and Lack, 1993; (ii) Gesco, 1993.

    C. (i) Moller and Leck, 1993; (ii) Gisco, 1993.

    D. (i) Maller and Lick, 1993; (ii) Gusco, 1993.

    E. (i) Meller and Lyck, 1993; (ii) Gisco, 1993.

11. What does the acronym TLF stand for?

    A. Total liquid fraction.

    B. Thin liquid film.

    C. Total liquid film.

    D. Thick liquid film.

    E. Transmembrane liquid force.

12. What effect does the Gibbs – Marangoni mechanism see manifested in alcoholic drinks or on surfaces concerned with wetting?

    A. Solubility—micellisation.

    B. Colour.

    C. Smell—rancidity.

    D. Tears and tear-ing—a pulling of liquid.

    E. Life-time.

13. Which of the following is *not* a medical – pharmaceutical colloid?

    A. A cream.

    B. A lotion.

    C. An interface.

D. A micelle.

E. A gel.

14. Which of the following should contain the least oil?

    A. Ointment.

    B. O/W cream.

    C. W/O cream.

    D. Paste.

    E. Aqueous lotion.

15. Drug exit from an emulsion droplet follows which type of diffusive process?

    A. Tyndallian.

    B. Lagrangian.

    C. Euclidean.

    D. Newtonian.

    E. Fickian.

16. How is the solubility constant, log P, defined?

    A. The square of the ratio of reactants divided by products.

    B. The ratio of (content in octanol) divided by (content in water).

    C. The ratio of reactants divided by products.

    D. A value between 0 and 1.

    E. The square root of the concentration (content in octanol) divided by (content in water).

17. Encapsulation of a drug usually involves what?

    A. Suspensions.

    B. Association colloids and emulsions.

    C. Micelles.

    D. Liposomes.

    E. Emulsions.

18. The 'best' stability assessment of coarse dispersions of medicines is usually assessed by which of the following?

    A. Drug content only.

    B. Rheology, visual appearance, drug content, microbiology.

    C. Accelerated testing only.

    D. Particle size.

    E. Visual appearance.

19. Benzoates E216 and E218 are used in formulated products as what?

    A. Emulsifiers.

    B. Adsorbents.

    C. Solubilisers.

    D. Preservatives, antimicrobials.

    E. Thickeners.

20. Ointments and creams are examples of what?

    A. Solids.

    B. Semisolids.

    C. Micelles.

    D. True solutions.

    E. Syrups.

21. Which is *not* a factor that influences emulsion stability?

    A. Mie theory.

    B. Polydispersity.

    C. Energetics of interaction.

    D. Flocculation.

    E. Kinetic driven flow.

22. What are the key ingredients in an emulsion?

    A. Vehicle, buffered media.

    B. Emulsifier, active.

    C. Emulsifier, active, vehicle, buffered media.

    D. Emulsifier, active, vehicle.

    E. Buffered media.

23. Drug dissolution rate from suspensions (e.g. micelles, nanoemulsions, microemulsions, emulsions) is related to which of the following?

    A. Drug solubility, layer thickness.

    B. Drug solubility, diffusion, surface area.

    C. Drug solubility, surface area, layer thickness.

    D. Drug solubility, diffusion, surface area, layer thickness.

    E. Drug solubility.

24. A high log P value indicates what?

    A. Apolarity of the drug.

    B. A weak acid drug.

    C. Neither polarity nor apolarity.

    D. Nothing; it is only related to buffering activity.

    E. A very polar drug species.

25. Which is the simplest form of viscosity behaviour and of material flow seen?

    A. Navier – Stokesian.

    B. Newtonian.

    C. Einsteinian.

    D. Leibnizian.

    E. Bernouillian.

26. Plastic rheological behaviour comes from what?

    A. Polyolefin chemistry.

    B. Separation of associated 'symmetric' particles in a flow field.

    C. Hydrolysis of asymmetric particles.

    D. Lipinski's rule-of-five theory.

    E. Elasticity associated with the covalent bonding between polymers forming a network.

27. What is the viscosity of very pure water (20 °C)?

    A. 85 mPas.

    B. 150 EPas.

    C. 1.00 mPas.

D. 8.4 µPas.

E. 0.2 kPas.

28. A pharmaceutical emulsion is 'normally' what?

A. A solid.

B. A solution.

C. A dispersion of liquid within another liquid.

D. A type of gel.

E. A type of gas.

29. Which phrase describes a microemulsion (or swollen micelle)?

A. A cluster of ionic micelles that can be used to solubilise impurities and drugs.

B. An agglomeration of lipid.

C. A collection of oil molecules.

D. Ten or so ions combined to form a solid lattice.

E. An aggregate of surfactant molecules that can be used to solubilise drugs in an oily core.

30. What is an emulsifier?

A. An amphiphile.

B. An acrophile.

C. An ambiphile.

D. A chemophile.

E. A thermophile.

Calculations

1. If the K_a of drug 'X' is 1.119×10^{-7}, what is the pK_a?

2. Using the Henderson – Hasselbalch equation, what is the exact pH of a drug if the pK_a is 1.7672 and the $\log_{10}$ ratio of salt to acid is 5.6021?

3. What is the pH of a $0.0511\,M$ HCl solution?

4. What is the pH of a $0.01916\,M$ HCl solution?

5. If the K_a of drug 'HA' is 1.178×10^{-6}, what is the pK_a?

6. What is the pK_a of an acid with K_as of 1.275×10^{-4} and 1.5×10^{-5}?

7. If the pK_a of a drug AH is 3.03, what is the K_a?

8. What is the pH of a $0.0062\,M$ KOH solution?

9. What is the total ratio of un-ionised drug (%) to ionised drug (%) if the $\log_{10}$ (base/acid) value is -0.50 and the pK_a and pH of drug HA are 7.00 and 6.50, respectively? *Assumptions*: simple acid model invoked; activity equals concentration; at the time of measurement, dissociation is complete; at the start, ([salt] + [acid]) is greater than [protons]. The compound is a weak acid, HA, with base $= [A^-]$ and acid $= $ [HA].

10. What would be the calculated log P of an acid drug, HA, with an oil concentration of $1.28 \times 10^{-2}\,M$ and an aqueous concentration of $1.55 \times 10^{-5}\,M$?

11. Is a drug with a log P value of 0.01 less or more water soluble than one with a value of 3.76?

12. What is the HLB of an EO/PO (E/P) polymer emulsifier, according to Griffin's original 1954 formula, HLB $= 20 \times (1 - (S/A))$, if the acid value is 40 and the saponification value is 8?

13. What is the HLB of an oxyethylene/polyhydric alcohol $((E + P)/5)$ emulsifier of e.g. a 10-carbon-chained sorbitol ester, if the number of oxyethylene units is 10 and the molecular mass of the sorbitol backbone is 183 Da? The molecular mass of EO is 44 Da, the percentage by weight of polyhydric alcohol is 13.9 and the total emulsifier molecular mass is 1.32 kDa.

14. What is the HLB of the product Surfactant-X, a 20 mol ethoxylate of oleyl alcohol? The ethylene oxide has a molecular weight of 44, oleic acid had a molecular weight of 270 and the species has a total molecular weight of 1150. Bear in mind HLB $= (E/5)$.

15. What is the HLB of a mix of two distinct emulsifiers, A and B, with HLB values of 4.9 and 15.5, respectively? The manufacturer uses a 60 : 40 blend of A : B.

16. What is the overall HLB if a scientist used 55% of emulsifier X, having an HLB of 12.6, and 45% of emulsifier Y, having an HLB of 11.7?

17. What would the spreading coefficient be if the water – air, oil – air and water – oil interfacial tensions were 52, 28 and 46 mN/m, respectively?

18. Use the Van't Hoff equation to find the sample purity if:

$$\frac{1}{F_s} = \Delta H \cdot \frac{(T_0 - T_s)}{RT_0{}^2} \cdot \frac{1}{X_2}$$

The values you have are as follows: sample temperature, $168\,^\circ\mathrm{C}$; pure substance melting point, $174\,^\circ\mathrm{C}$; enthalpy, $28.1\,\mathrm{kJ/mol}$. The molar fraction of impurity in the sample is unknown, but the gas constant is $8.314\,\mathrm{J/K/mol}$ and the fraction of sample melted at sample temperature is 0.02986. T_s is the sample temperature, T_0 is the pure substance melting point, ΔH is the melting enthalpy, X_2 is the molar fraction of impurity in the sample, R is the gas constant, F_s is the fraction of sample melted at T_s (this is equal to the area of endotherm to T_s/total endotherm area (refer to Figure 12.4, insets)). *Assumption*: fully soluble solutes (eutectic mixture) at thermodynamic equilibrium; no sample decomposition and the concentration of impurity is low. (7 marks)

19. What would be the calculated enthalpy (kJ/mg) and would the reaction be endothermic or exothermic when:

$$A = -kGm\,\Delta H$$

The analyst looks at the thermogram produced by the DSC, which uses $5.1\,\mathrm{mg}$ of drug 'X', and the peak area is $+357.5\,\mathrm{mJ}$. The geometric calibration factor is 22.5 and the thermal conductivity constant is $4\,\mathrm{J/s/m/K}$. *Considerations*: if using the same equipment, the expression kG becomes a constant. A is the peak area, k is the thermal conductivity constant, m is the mass of the sample, G is the geometric calibration factor and ΔH is the enthalpy (in Joules per unit amount). (3 marks)

20. If it takes a robot arm 3.2 minutes to sample a 96-well plate $(= 12 \times 8$ individual wells), how long in minutes would it take it to sample a 364 and a 1536 well plate, respectively? Why would such a large number of samples be examined simultaneously and for what purpose? (5 marks)

21. The Langmuir trough is used for examining insoluble monolayers of surface-active water-insoluble pharmaceutical materials. Assuming Avogadro's number $(N_a = 6.022 \times 10^{23}\,\mathrm{mol^{-1}})$ of molecules equals a mole, what would the cross-sectional area per molecule in square Angstroms (A_{mol}) be in $\mathrm{nm^2}$ for an insoluble monolayer of distearoyl phosphatidylcholine (DSPC) lipid on the surface of buffered water at ambient room temperature if k_{expt} is 2×10^{-16} and Γ is $1.1 \times 10^{-9}\,\mathrm{mol/cm^2}$? You may use the following equation:

$$A_{mol} = (k_{expt})/(\Gamma \times N_a)$$

Assumptions: zero solubility of surfactant in solvent-support media. The volatile solvent evaporates instantaneously on deposition on the surface. The sample at measurement is at *equilibrium*. The shape of the molecule is modelled on a cylinder with base on the interface, the base being solvated by water molecules. k_{expt} is the experimental constant for concentration of the gas constant and surface area and Γ is the molar excess concentration in mol/cm^2. (5 marks)

22. If the diameter of a spherical SLN fat particle is 25 nm, what is the difference in volume if the bare particle takes on an absorbed poly(ethylene glycol) (PEG) polymer layer of average depth 15 nm? The volume of a sphere is 4/3 πr^3. (5 marks)

Dilemmas

Consider the following dilemmas and polemical topics, which have no absolute answer. What would you propose based on reading this book?

1. Why is there a constant need for novel new blockbuster drugs? (Think about a square peg in a round hole.)

2. Better drugs: does this mean prophylactics or palliatives? (Think about quality of life versus curatives.)

3. Is there such a thing as a 'side effect-free' drug and what would the drug developer's perspective be? (Think about fewer consequences and complementarity with a drug with marked side effects.)

4. Why is competitive advantage important to both the drug company and the patient? (think about a technology push and innovation.)

5. Why does it profit a regulatory body to have more new medicine and drug delivery forms? (Think about the patient's and clinician's needs.)

6. Why is there a need for weightiness of data in a clinical trial and thus in a product licence application?

7. Development of new drugs has been said to provide a 'chance of true good'. Would you say this is true or false, and why?

8. In the development of a new dosage form, does the company (ethically) have to declare all findings? If so, why, and if not, why not?

9. Statistics can be used to prove drug product efficacy. Why is this important?

10. Why are societal pressure, company complacency, coercion, rigour of scientific method and potential data 'massaging' so important and inhibitory to sound *de novo* and innovative drug development?

11. Why is drug product form so important to its efficacy *in vivo*?

12. If you were asked to describe the most significant drug dosage form in modern pharmaceutics, what would it be and why?

13. If you, as a process and product developer, were asked to pick the six most important analytical methods that support drug product QA/QC and development, what would they be and why?

14. Why are surface and bulk chemistry and physics so import to almost every conceivable 'emulsion'-type pharmaceutical product?

15. Where do you see the emphasis of future drug development in connection with emulsion products coming from and why?

16. Why would a manufacturer choose to make smaller particles for drug delivery?

17. Why are emulsion products generally not used for oral drug delivery?

18. Which is the more important entropy, enthalpy or free energy in the steric stabilisation mechanism implicit in 'DLVO' theory?

19. How do surfactants function to lower the interfacial energy?

20. Why does temperature increase emulsion instability?

Answers

Short

1. Less net pull at interface between adjacent molecules – excess surface energy, $+\Delta G$. Manifested in a 'toughening' of the surface.

2. Line tensions (γ) of LV, SL, LS and surface roughness (surface structuring), viscosity (assumes planar horizontal surface at STP). Time of study and duration/evaporation from the drop may also feature.

3. Babraham Institute, 1961 – 1964. Never won Nobel Prize, Taxol, Caelyx, Doxil products based on liposome. SUV to MLV. All forms of encapsulation (self-assembly) with variation in amphiphiles or lipid composition (and sketch).

4. Poorer water solubility (this is better), therefore make association structures with bilayers. Chemical stability enhanced, variation in amphiphiles or lipid composition (and sketch). Temperature-driven and/or electromagnetic release of liposomal encapsulated drug (SPIO metal particles). For analysis/diagnosis or therapy.

5. Qutenza (capsaicin in SLN in patch, analgesia/neuralgia). Gel phase lipid with significant viscosity, thus control over diffusion/flux/efflux, entrapment

and solubility. Radioactive ^{111}In (indium), ^{99m}Tc (technetium), ^{153}Gd (gadolinium) DPTA inclusion for MRI contrast agents and therapeutics based on the specific metabolism of substrates bearing these entities.

6. Abraxane (FDA approved 2010) (paclitaxel-HSA; cancer), DermaVir (plasmid DNA (for AIDS/HIV), in polymeric micelle), Qutenza (capsaicin in SLN in patch, analgesia/neuralgia). All forms of encapsulation (self-assembly) with variation in amphiphiles or lipid composition (and sketch).

7. Poorer water solubility (this is better), therefore make association structures with bilayers.

8. DermaVir (plasmid DNA (for HIV/AIDS), in polymeric micelle), Abraxane (paclitaxel-HAS conjugate; cancer), Qutenza (capsaicin in SLN in patch, pain relief/neuralgia). All forms of encapsulation (self-assembly) with variation in amphiphiles or lipid/lipid derivative composition (with sketch)

9. Entanglement (reptation), interaction, egg – box junction (ionic bridge, bifunctional agent). Mixture.

10. Coalescence (surface rheology), disproportionation/Ostwald ripening (solid mixture combined with surface coverage to saturation); thermomechanical agitation/perturbation (surface pressure, adsorbed particulates, lateral diffusion, water/solvent binding).

11. Microemulsions, biofilms, multiple emulsions, emulsions, sols, liposomes, liquid crystals, gels, transdermal patches, SLNs/SLCs/LNPs/NPs, colloidal polymers (peptides), micelles, foams (liquid, solid, dry), combinations of these.

12. Contribution = repulsive (charge) + attractive (polarisation) + steric (surface 'shape' or 'coating'). ΔG, ΔH and ΔS important. Four scientists (named): 'DLVO'. Basic form of plot (sketch) – potential versus distance. Relevance to pharmacy: many weakly acidic/basic (ionised), therefore charged or poorly water-soluble drugs. Examples = aspirin, barbiturates, amino acids, penicillin. Product stability – how implicated? Examples of consequences: flocculation, coalescence.

13. O/W, W/O, multiple (W/O/W; O/W/O), micro- O/W and W/O (need cosurfactant). Surfactant/emulsifier/polymer (even particles) use stabilise the droplet surface by lowering the surface/interfacial tension (sketch). Dispersed and continuous phases – means what?, mixing/shear. MEM is 100 nm, other emulsions $\sim$1 – 100 microns. Quality: creaming, sedimentation, coalescence, flocculation, splitting/inversion, chemical change. Size = $kT/6\pi\eta D$. Creaming – viscosity, density, size, coalescence, flocculation - DLVO, surface rheology-surfactant types.

Long

1. DSC, melting point and crystalline/amorphous, polymorph detection, purity/crystallinity quantification, excipients compatibility, water measurement, qualitative comparison. Strengths: cheap (£50 – 100k), specialist, diverse (kinetics to gel – sol transition) use. Weaknesses: awkward, 'unpopular', not easily hyphenated. Only technique for polymorphs and pseudopolymorphs or monotropism/enantiotropism. Essential in preformulation of drugs/excipients, e.g. sulphonamides, barbiturates, steroids, fats, peptides.

2. Emulsifier, vehicle, thickener, buffers, preservative, 'active'. Hygiene, PCQ, validation, in-process control, batch control, specifications, HACCP.

3. NEITHER all impact on PCQ. Futility in proceeding in terms of validation, in-process control, batch control, specifications, HACCP.

4. Tight control over nonconformance driven by a rigorous and defined process achieved by validation and SPC.

5. Risk in making a faulty product. Process relies on initial validation. Benchmarks for product specifications. Dedicated facilities/equipment used. QP involvement per batch. Batch documents and SOPs at all stages. Index tests for 'quality' are key, e.g. drug release profile. Form of product mapped. Consistency, purity and quality (hygiene) of product controlled. Known process variables (sets standards). Use what learnt for next campaign or production run (validation). Track number and type of noncompliances.

6. Sterility, predictability, PCQ, safety – quality – efficacy. Risk in making a faulty product.

7. Toxicology, specificity, suitability. Solubility, log P, pK_a, environmental sensitivity (light, pH, redox). Advised route of administration. Need preformulation and appropriate mixing and use of excipients.

8. MI^2 (see Sarker, 2008). Specifics of form. GCP/GMP/GLP. Lead-NCE-IND-NDA-AND (approved new drug). Customer and business requirements. Control of quality/consistency (PCQ). Legal status of new product, PL. Answer must include worked examples of pharmaceutical type.

MCQs

1C, 2A, 3E, 4B, 5B, 6E, 7A, 8C, 9E, 10A, 11B, 12D, 13C, 14E, 15E, 16B, 17B, 18B, 19D, 20B, 21A, 22C, 23D, 24A, 25B, 26B, 27C, 28C, 29E, 30A.

Calculations

1. 6.951.

2. $pH = pK_a + \log (\text{salt/acid}) = 7.3693$.

3. 1.292, assumes strong acid.

4. 1.72.

5. 5.93.

6. $(1.275 \times 10^{-4} + 1.5 \times 10^{-5})/2 = 4.355$.

7. 9.33×10^{-4}.

8. $\text{Log } 0.0062 = x$; $pK_w = (pH + pOH) = 14$; $14 - x = 11.79$. Assumes strong base.

9. Henderson – Hasselbalch equation. $6.5 - 7.0 = \log{-0.5}$; antilog $-0.5 = b/a = 0.316$; $0.316 \, A-$ for each HA; total amount $= HA + A- = (0.316 + 1)$; fraction therefore $= (1/1.316)$; un-ionised $= 76\%$; ratio un-ionised to ionised $= 76 : 24\%$.

10. $\text{Log [oil]} : \text{[water]} = \log (1.28 \times 10^{-2} \, M/1.55 \times 10^{-5} \, M) = 2.917$.

11. Antilog 0.01 and 3.76; $3715 \gg 1.02$, so the higher indicates poorer water solubility.

12. $HLB = 20 \times (1 - (8/40)) = 20 \times 0.8 = 16$.

13. $HLB = ((E + P)/5)$
$E = (10 \times 44 \times 100)/1320 = 33.3\%$
$P = (183 \times 100)/1300 = 13.9\%$
$33.3 + 13.9 = 47.2\%$
$HLB = 47.2/5 = 9.4$.

14. $(20 \times 44) \, 880$
$880 + 270 = 1150$
$(880/1150) \times 100 = 76.52\%$
$HLB = 76.52/5 = 15.3$.

15. $(0.6 \times 4.9) + (0.4 \times 15.5) = 2.94 + 6.2$; $HLB = 9.1$.

16. $(0.55 \times 12.6) + (0.45 \times 11.7) = 6.93 + 5.27$; $HLB = 12.2$.

17. $S = 28 - (58 + 46) = -76$, value is very negative, spreading does not occur.

18. $(1/0.02986) = \{(28\,100 \, \text{J}/8.314) \times [(174 - 168)/(174)^2] \times 1/\text{unknown}\}$
$1/0.02\,986 = 33.4896 = 3378.9 \times (1.982 \times 10^{-4}) \times 1/X_2$. Rearrange for X_2, so answer $= 0.02$ or $1/50$.

19. $kG \rightarrow 1.\Delta H = A/m$, so $= 357.5/5.1 = 70.098$ mJ/mg. Endothermic (+ve ΔH), like all melting processes.

20. $3.2 \times 60 = 192$ seconds. $192/96$ wells $= 2$ seconds per well. So 386 wells $= 772$ seconds (12.87 minutes) and 1536 wells $= 3072$ seconds (51.2 minutes). Large for high-throughput screening, e.g. cell or pharmacological studies involving many repeats of multiple tests.

21. $(2 \times 10^{16})/(1.1 \times 10^{-9} \times 6.022 \times 10^{23}) = 30.19$; divide area by $(10 \times 10 = 100)$ to get in nm^2, so answer $= 30.19/100 = 0.3\,nm^2$.

22. $25/2 = $ radius $= 12.5$; $12.5 + 15 = 27.5$ nm new radius;
so $1.333 \times 3.14 \times (12.5)^3 = 8175\,nm^3$; then:
$1.333 \times 3.14 \times (27.5)^3 = 87048\,nm^3$
$(87048\,nm^3/8175\,nm^3) \times 100 = 1065\%$ increase in particle volume.

Dilemmas

Answers to be found within the textbook.

References

Acartürk, F. (2009) *Recent Patents on Drug Delivery & Formulation*, **3**: 193–205.

Adamson, A.W. (1990) *Physical Chemistry of Surfaces*, John Wiley & Sons Ltd, New York.

Alayoubi, A., Kanthala, S., Satyanaranajois, S.D., Anderson, J.F., Sylvester, P.W. and Nazzal, S. (2013) *Colloids and Surfaces B: Biointerfaces*, **103**: 23–30.

Al-Hanbali, O., Rutt, K.J., Sarker, D.K., Hunter, A.C. and Moghimi, S.M. (2006) *Journal of Nanoscience and Nanotechnology*, **6**(8): 3126–3133.

Almeida, A.J. and Souto, E. (2007) *Advanced Drug Delivery Reviews*, **59**: 478–490.

An, H.Z., Hegelson, M.E. and Doyle, P.S. (2012) *Advanced Materials*, **24**: 3838–3844.

Ansel, H.C., Allen, L.V. and Popovich, N.G. (1999) *Pharmaceutical Dosage Form and Drug Delivery*, Lippincott Williams & Wilkins, Philadelphia.

Araujo S.C., Mattos, A.C.A., Teixeira, H.F., Coelho, P.M.Z., Nelson, D.L. and Oliveira, M.C. (2007) *International Journal of Pharmaceutics*, **337**: 307–315.

Araujo, F.A., Kelmann, R.G., Araujo, B.V., Finatto, R.B., Teixeira, H.F. and Koester, L.S. (2011) *European Journal of Pharmaceutical Sciences*, **42**: 238–245.

Arditty, S., Schmitt, V., Giermanska-Kahn, J. and Leal-Calderon, F. (2004) *Journal of Colloid and Interface Science*, **275**: 659–664.

Aulton, M.E. (2002) *Pharmaceutics: The Science of Dosage Form Design*, Churchill-Livingstone, London.

Aveyard, R., Binks, B.P. and Clint, J.H. (2003) *Advances in Colloid and Interface Science*, **100–102**: 503–546.

Bachhav, Y.G. and Patravale, V.B. (2009) *International Journal of Pharmaceutics*, **365**: 175–179.

Bangham, A.D. and Horne, R.W. (1964) *Journal of Molecular Biology*, **8**: 660–668.

Becher, P. (2001) *Emulsions: Theory and Practice*, 3rd edition, Oxford University Press, Oxford.

Benoliel, M.J. (1999) *Trends in Analytical Chemistry*, **18**: 632–638.

Benson, H.A. and Watkinson, A.C. (2012) *Topical and Transdermal Drug Delivery: Principles and Practice*, John Wiley & Sons Ltd, Boca Raton.

Bhalekar, M.R., Pokharkar, V., Madgulkar, A., Patil, N. and Patil, N. (2009) *AAPS PharmSciTech*, **10**: 289–296.

Binks, B.P. (2002) *Current Opinion in Colloid & Interface Science*, **7**: 21–41.

Bivas-Benita, M., Oudshoorn, M., Romeijn, S., Miejgaarden, K., Koerten, H., Meulen, H., Lambert, G., Ottenhoff, T., Benita, S., Junginger, H. and Borchard, G. (2004) *Journal of Controlled Release*, **100**: 145–155.

Boerman, O.C., Oyen, W.J.G., Storm, G., Corvo, M.L., van Bloois L., van der Meer, J.W.M. and Corstens, F.H.M. (1997) *Annals of the Rheumatic Diseases*, **56**: 369–373.

Boevski, I., Genov, K., Boevska, N., Milenova, K., Batakliev, T., Georgiev, V., Nikolov, P. and Sarker, D.K. (2011) *Comptes rendus de l'Académie bulgare des Sciences*, **64**(1): 33–38.

Brusewitz, C., Schendler, A., Funke, A., Wagner, T. and Lipp. R. (2007) *International Journal of Pharmaceutics*, **329**: 173–181.

Busse, M.J. (1978) *Pharmaceutical Journal*, **220**: 25–26.

Capelle, M.A.H., Gurny, R. and Arvinte, T. (2007) *European Journal of Pharmaceutics and Biopharmaceutics*, **65**: 131–148.

Castelletto, V., Cantat, I., Sarker, D., Bausch, R., Bonn, D. and Meunier, J. (2003) *Physical Review Letters*, **90** (4): 048302.

Cavalli, R., Gasco, M. R., Chetoni, P., Burgalassi, S. and Saettone, M.F. (2002) *International Journal of Pharmaceutics*, **238**: 241–245.

Cevc, G. and Vierl, U. (2010) *Journal of Controlled Release*, **141**: 277–299.

Chaiseri, S. and Dimick, P.S. (1986) *Journal of the American Oil Chemists' Society*, **72**(12): 1491–1496.

Chen, H., Chang, X., Du, D., Liu, W., Liu, J., Weng, T., Yang, Y., Xu, H. and Yang, X. (2006) *Journal of Controlled Release*, **110**: 296–306.

Chen, J., Dickinson, E. and Iveson, G. (1993) *Food Structure*, **12**: 135–146.

Choi, S.-W., Zhang, Y. and Xia, Y. (2010) *Angewandte Chemie International Edition (English)*, **49**(43): 7904–7908.

Clark, D.C., Wilde, P.J. and Wilson, D.R. (1991a) *Colloids and Surfaces*, **59**: 209–223.

Clark, D.C., Dann, R., Mackie, A.R., Mingins, J., Pinder, A.C., Purdy, P.W., Russell, E.J., Smith, L.J. and Wilson, D.R. (1991b) *Journal of Colloid and Interface Science*, **138**: 195–206.

Collins, G., Patel, A., Dilley, A. and Sarker, D.K. (2008) *Journal of Agricultural Food Chemistry*, **56**(10): 3846–3855.

Concannon, C., Hennelly, D.A., Noott, S. and Sarker, D.K. (2010) *Current Drug Discovery Technologies*, **7**: 123–136.

Constantinides, P.P. (1995) *Pharmaceutical Research*, **12**: 156–162.

Conzen, P.F. (2005) *Best Practice & Research Clinical Anaesthesiology*, **17**(1): 29–46.

Corveleyn, S. and Remon, J.P. (1998) *International Journal of Pharmaceutics*, **166**(1): 65–74.

Courier, H.M., Pons, F., Lessinger, J.M., Frossard, N., Krafft, M.P. and Vandamme, T.F. (2004) *International Journal of Pharmaceutics*, **282**: 131–140.

Courthaudon, J.-L., Dickinson, E. and Matsumura, Y. (1991) *Colloids and Surfaces*, **56**: 293–300.

Csaba, N., Garcia-Fuentes, M. and Alonsa, M.J. (2009) *Advaned Drug Delivery Reviews*, **61**: 140–157.

D'Ascenzo, R., D'Egidio, S., Angelini, M.P., Mannam, M., Pompillo, A., Cogo, P.E. and Carnielli, V.P. (2011) *Journal of Pediatrics*, **159**(1): 33–38.

Davis, M.E., Chen, Z.G. and Shin, D.M. (2008) *National Review of Drug Discovery*, **7**: 771–782.

Desi Reddy, R.B., Kumari, C.T.L., Sowjanya, G.N., Sindhuri, S.L. and Bandhavi, P. (2012) *International Journal of Pharmaceutical Research and Development*, **4**: 137–152.

Di Mattia, C.D., Sacchetti, G., Mastrocola, D., Sarker, D.K. and Pittia, P. (2010) *Food Hydrocolloids*, **24**: 652–658.

Dickinson, E. and Hong, S.-T. (1994) *Journal of Agricultural Food Chemistry*, **42**: 1602–1606.

Dimitrova, T.D., Gurkov, T.D., Vasssileva, N., Campbell, B. and Borwankar, R.P. (2000) *Journal of Colloid and Interface Science*, **230**: 254–267.

Dinsmore, A.D., Hsu, M.F., Nikolaides, M.G., Marquez, M., Bausch, A.R. and Weitz D.A. (2002) *Science*, **298**: 1006–1009.

Eldem, T., Speiser, P. and Hinkal, A. (1991) *Pharmaceutical Research*, **8**: 47–54.

Fang, J.Y., Leu, Y.L., Chang, C.C., Lin, C.H. and Tsai, Y.H. (2004) *Drug Delivery*, **11**: 97–105.

Fitzpatrick, S.D., Fitzpatrick, L.E., Thakur, A., Mazumder, M.A. and Sheardown, H. (2012) *Expert Reviews of Medical Devices*, **9**(4): 339–351.

Florence, A.T. and Attwood, D. (1998) *Physicochemical Principles of Pharmacy*, Macmillan Press, London.

Freitas, C. and Müller, R.H. (1999) *European Journal of Pharmaceutics and Biopharmaceutics*, **47**: 125–132.

Gasco, M.R. (1993) Method for producing solid lipid nanospheres having a narrow distribution (Italy), US patent 188837.

Gelperina, S., Kisich, K., Iseman, M. D. and Heifets, L. (2005) *American Journal of Respiratory and Critical Care Medicine*, **172**: 1487–1490.

Genov, K., Boevska, N., Boevski, I. and Sarker, D.K. (2011) *Comptes rendus de l'Académie bulgare des Sciences*, **64**(4): 509–514.

Georgiev, G.A., Sarker, D.K., Al-Hanbali, O., Georgiev, G.D. and Lalchev Z. (2007) *Colloids and Surfaces B: Biointerfaces*, **59**: 184–193.

Gerweck, L.E., Vijayappen, S. and Kozin, S. (2006) *Molecular Cancer Therapy*, **5**(5): 1275–1279.

Ghose, A.K., Viswanadhan, V.N. and Wendoloski, J.J. (1999) *Journal of Combinatorial Chemistry*, **1**: 55–68.

Gianella, A., Jarzyna, P.A., Mani, V., Ramachandran, S., Calcagno, C., Tang, J., Kann, B., Dijk W.J., Thijssen, V.L., Griffioen, A.W., Storm, G., Fayad, Z.A. and Mulder, W.J. (2011) *ACS Nano*, **5**(6): 4422–4433.

Goodwin, J.W. (2000) *Rheology for Chemists: An Introduction*, RSC Publishing, Cambridge.

Grammen, C., Augustijns, P. and Brouwers, J. (2012) *Antiviral Research*, **96**: 226–233.

Gregoriadis, G. (1973) *FEBS Letters*, **36**: 292–296.

Gregoriadis, G. (1977) *Nature*, **265**(3): 407–411.

Guo, Y., Liu, X., Sun, X., Zhang, Q., Gong, T. and Zhang, Z. (2012) *Theranostics*, **2**(11): 1104–1114.

Hansen, T., Holm, P., Rohde, M. and Schultz, K. (2005) *International Journal of Pharmaceutics*, **293**(1–2): 203–211.

Hayes, M.E., Drummond, D.C. and Kirpotin, D.B. (2006) *Gene Therapy*, **13**: 646–651.

Hegelson, M.E., Moran, S.E., An, H.Z. and Doyle, P.S. (2012) *Nature Materials*, **11**: 344–352.

Henzl, M.R. (2005) *American Journal of Drug Delivery*, **3**: 227–237.

Hiemenz, P.C. and Rajagopalan, R. (1997) *Principles of Colloid and Surface Chemistry*, Marcel Dekker, New York.

Horne, R.W., Bangham, A.D. and Whittaker, V.P. (1963) *Nature*, **200**: 1340.

Howbrook, D., Sarker, D., Lloyd, A.W. and Louwrier, A. (2002) *Biotechnology Letters*, **24**: 2071–2074.

Howbrook, D.N., van der Valk, A.M., O'Shaughnessy, M.C., Sarker, D.K., Baker, S.C. and Lloyd, A.L (2003) *Drug Discovery Today*, **8** (14): 642–651.

Hu, X. and Dahl, G. (1999) *FEBS Letters*, **451**: 113–117.

Huailiang, W.U., Ramachandran, C., Bielinska, A.U, Kingzett, K., Sun, R., Weiner, N.D. and Roessler, B.J. (2001) *International Journal of Pharmaceutics*, **221**: 23–34.

Jain A., Agarwal A., Majumder S., Lariya, N., Khaya, A., Agrawal, H., Majumdar, S. and Agrawal, G.P. (2010) *Journal of Controlled Release*, **148**(3): 359–367.

Jee, J.P., Parlato, M.C., Perkins, M.G., Mecozzi, S. and Pearce, R.A. (2012) *Anesthesiology*, **116**(3): 580–585.

Jones, K. A. and Harmanli, O. (2010) *Reviews in Obstetrics and Gynecology*, **3**, 3–9.

Jones, R.A.L. (2002) *Soft Condensed Matter*, Oxford University Press, Oxford.

Jordan, M., Nayel, A., Brownlow, B. and Elbayoumi, T. (2012) *Journal of Biomedical Nanotechnology*, **8**: 944–956.

Kaur, I.P., Bhandari, R., Bhandari, S. and Kakkar, V. (2008) *Journal of Controlled Release*, **127**: 97–109.

Kazi, K.M., Mandal, A.S., Biswas, N., Guha, A., Chatterjee, S., Behera, M. and Kuotsu, K. (2010) *Journal of Advanced Phamaceutical Technology & Research*, **1**(4): 374–380.

Krafft, M.P. and Riess, J.G. (2009) *Chemical Reviews*, **109**: 1714–1792.

Kumar, M., Misra, A., Babbar A.K. and Mishra A.K. (2008) *International Journal of Pharmaceutics*, **358**: 285–291.

Lawrence, M.J. and Rees, G.D. (2012) *Advanced Drug Delivery Reviews*, **64**: 175–193.

Lemke, T.L., Williams, D.A., Roche, V.F. and Zito, S.W. (2013) *Foye's Principles of Medicinal Chemistry*, 7th edition, Lippincott Williams & Wilkins, Philadelphia.

Lipinski, C.A. (2000) *Journal of Pharmaceutical and Toxicological Methods*, **44**: 235–249.

Lipinski, C.A., Lombardo, F., Dominy, B.W., Feeney, P.J. (1997) *Advanced Drug Delivery Reviews*, **23**: 3–25.

Lui, K.C. and Chow, Y.F. (2010) *Hong Kong Medical Journal*, **16**(6): 470–475.

Lupo, M.P. (2001) *Clinical Dermatology*, **19**: 467–473.

Maeda, S., Nakagawa, S., Suga, M., Yamashita, E., Oshima, A., Fujiyoshi, Y. and Tsukihara, T. (2009) *Nature*, **458**(7238): 597–602.

Mahato, R.I. (2007) *Pharmaceutical Dosage Forms and Drug Delivery*, CRC Press, Boca Raton.

Manoj, P., Fillery-Travis, A.J., Watson, A.D., Hibberd, D.J. and Robbins, M.M. (1998) *Journal of Colloid and Interface Science*, **207**: 283–293.

Mansour, H.M., Rhee, Y.-S. and Wu, X. (2009) *International Journal of Nanomedicine*, **4**: 299–319.

Marangoni, C.G.M. (1871) *Annalen Physik Chemie (Poggendorff)*, **143**(7): 337–354.

Maurer, N., Fenske, D.B. and Cullis, P.R. (2001) *Expert Opinion on Biological Therapy*, **1**: 1–25.

Mehnart, W. and Mader, K. (2001) *Advanced Drug Delivery Reviews*, **47**: 165–196.

Mei, Z. and Wu, Q. (2005) *Drug Development and Industrial Pharmacy*, **31**: 161–168.

Moghimi, H.R. (1996) *International Journal of Pharmaceutics*, **131**: 117–120.

Moghimi, S.M., Hunter, A.C. and Murray, J.C. (2005) *FASEB*, **19**: 311–330.

Moses, M.A., Brem, H. and Langer, R. (2003) *Cancer Cell*, **5**: 337–341.

Müller, R.H. and Lucks, J.S. (1996) Medication vehicles made of solid lipid nanospheres (Germany), European patent 0605497.

Müller, R.H. and Böhm, B.H.L. (1998) In: Müller, R.H., Benita, S. and Böhm, B. (Eds) *Emulsions and Nanosuspensions for the Formulation of Poorly Soluble Drugs*, Medpharm GmbH Scientific Publishers, Stuttgart, pp. 151–160.

Müller, R.H., Mehnert, W., Lucks, J.S., Schwarz, C., zur Mühlen, A., Weyhers, H., Freitas, C. and Ruhl, D. (1995) *European Journal of Pharmaceutics and Biopharmaceutics*, **41**(1): 62–69.

Müller, R.H., Mäder, K., and Gohla S. (2000) *European Journal of Pharmaceutics and Biopharmaceutics*, **50**: 161–177.

Muller, K.M., Gempeler, M.R., Scheiwe, M.-W. and Zeugin, B.T. (1996) *Pharmaceutica Acta Helvetiae*, **71**(6): 421–438.

Nikolov P., Genov, K., Konova, P., Milenova, K., Batakliev, T., Georgiev, V., Kumar, N., Sarker, D.K., Pishev, D. and Rakovsky, S. (2010) *Journal of Hazardous Materials*, **184**: 16–19.

Nishioka, B., Watanabe, S., Fijita, Y., Kojima, O., Morisawa, K., Yamane, E., Umehara, M. and Majima, S. (1980) *Japanese Journal of Surgery*, **10**: 110–114.

Noguchi, T., Takahashi ,C., Kimura T., Muranishi, S. and Sezaki, H. (1975) *Chemical & Pharmaceutical Bulletin*, **23**: 775–781.

Pashley, R.M. and Karaman, M.E. (2004) *Applied Colloid and Surface Chemistry*, John Wiley & Sons Ltd, Chichester.

Pickering, S.U. (1907) *Journal of the Chemical Society*, **91**: 2001–2021.

Popp, S.M. (2009) Pharmaceutical dosage forms fabricated with nanomaterials for quality monitoring (USA), US patent 20090004231.

Porter, C.J., Moghimi, S.M., Illum L. and Davis, S.S. (1992) *FEBS Letters*, **305**: 63–66.

Pouton, C.W. (2000) *European Journal of Pharmaceutical Sciences*, **11**: S93–98.

Primo, F.L., Michieloto, L., Rodrigues, M.A.M., Macaroff, P.P, Morais, P.C, Lacava, Z.G.M., Bently, M.V.L.B and Tedesco, A.C. (2007) *Journal of Magnetism and Magnetic Materials*, **311**: 354–357.

Qian, Y., Zha, Y., Feng, B., Pang, Z., Zhang, B., Sun, X., Ren, J., Zhang, C., Shao, X., Zhang, Q. and Jiang, X. (2013) *Biomaterials*, **34**: 2117–2129.

Rafai, S., Sarker, D. Bergeron, V., Meunier, J. and Bonn, D. (2002) *Langmuir*, **18**: 10 486–10 488.

Ramsden, W. (1903) *Proceedings of the Royal Society of London*, **72**: 156–164.

Rao, K.M., Mallikarjuna, B., Rao, K.S.V.K., Siraj, S., Rao, K.C. and Subha, M.C.S. (2013) *Colloids and Surfaces B: Biointerfaces*, **102**: 891–897.

Richards, R.L., Rao, M., Vancott, T.C., Matyas, G.R., Birx, D.L. and Alving, C.R. (2004) *Immunology & Cell Biology*, **82**: 531–538.

Riess, J.G. and Krafft, M.P. (1998) *Biochimie*, **80**: 489–514.

Roberts, R.J. (1981) *Pediatric Clinics of North America*, **28**: 23–34.

Robieux, I., Kumar, R., Radhakrishnan, S. and Koren, G. (1991) *Journal of Pediatrics*, **118**(6): 971–973.

Ruan, W., French, D., Wong, A., Drasner, K. and Wu, A.H. (2012) *Anesthesiology*, **116**(2): 334–339.

Rudolph, C., Schillinger, U., Ortiz, A., Tabatt, K., Plank, C., Muller, R.H. and Rosenecker, J. (2004) *Pharmaceutical Research*, **21**(9): 1662–1669.

Russell, A.D. (1991) *Journal of Applied Bacteriology*, **71**: 191–201.

Sarker, D.K. (2002) *Journal of Pharmacy and Pharmacology*, **54**: S2.

Sarker, D.K. (2004a) *Drug Discovery Today*, **9**(2): 95.

Sarker, D.K. (2004b) *Business Briefing: Future Drug Discovery*, **1**: 38–41.

Sarker, D.K. (2005a) *Current Drug Delivery*, **2**(4): 297–310.

Sarker, D.K. (2005b) *Current Nanoscience*, **1**(2): 157–168.

Sarker, D.K. (2006a) *Mini-reviews in Medicinal Chemistry*, **6**(7): 793–804.

Sarker, D. (2006b) *PFQ (Pharmaceutical Formulation & Quality)*, **8**(5): 42–46.

Sarker, D.K. (2008) *Quality Systems and Controls for Pharmaceuticals*, John Wiley & Sons Ltd, Chichester.

Sarker, D.K. (2009a) *Current Drug Discovery Technologies*, **6**(1): 52–58.

Sarker, D.K. (2009b) In: Li, S. (Ed.) *Current Focus on Colloids and Surfaces*, Transworld Research Network, Kerala, pp. 225–242.

Sarker, D.K. (2010) *Recent Patents in Material Science*, **3**(3): 191–202.

Sarker, D.K. (2012a) In: Tiwari, A., Ramalingam, M., Kobayashi, H. and Turner, A.P.F. (Eds) *Biomedical Materials and Diagnostic Devices*, Scrivener-Wiley, Beverly, pp. 395–434.

Sarker, D.K. (2012b) In: *Transitions: Quality Adaptability and Sustainability in Times of Change*, Centre for Learning and Teaching Publications, University of Brighton, UK, pp. 63–76.

Sarker, D.K. and Wilde, P.J. (1999) *Colloids Surfaces B: Biointerfaces*, **15**: 203–213.

Sarker, D.K., Wilde, P.J. and Clark, D.C. (1995a) *Colloids Surfaces B: Biointerfaces*, **3**: 349–356.

Sarker, D.K., Wilde, P.J. and Clark, D.C. (1995b) *Journal of Agricultural Food Chemistry*, **43**: 295–300.

Sarker, D.K., Wilde, P.J. and Clark, D.C. (1996) *Colloids Surfaces A: Physicochemical Engineering Aspects*, **114**: 227–236.

Sarker, D.K., Wilde, P.J. and Clark, D.C. (1998a) *Cereal Chemistry*, **75**(4): 493–499.

Sarker, D.K., Bertrand, D., Chtioui, Y. and Popineau, Y. (1998b) *Journal of Texture Studies*, **29**: 15–42.

Sarker, D.K., Axelos, M. and Popineau, Y. (1999) *Colloids Surfaces B: Biointerfaces*, **12**: 147–160.

Seelig, A. (2007) *Journal of Molecular Neuroscience*, **33**: 32–41.

Sinko, P.J. (2006) *Martin's Physical Pharmacy and Pharmaceutical Sciences*, 5th edition, Lippincott Williams & Wilkins, Philadelphia.

Sonavanea, G., Tomodaa, K. and Makinoa, K. (2008) *Colloids Surfaces B: Biointerfaces*, **66**: 274–280.

Souto, E.B. and Muller, R.H. (2005) *Journal of Microencapsulation*, **22**: 501–510.

Souto, E.B., Wissing, S.A., Barbosa, C.M. and Muller, R.H. (2004) *International Journal of Pharmaceutics*, **278**: 71–77.

Suzuki, A., Morishita, M., Kajita, M., Takayama, K., Isowa, K., Chiba, Y., Tokiwa, S. and Nagai, T. (1998) *Journal of Pharmaceutical Sciences*, **87**: 1196–1202.

Tang, S.Y., Sivakumar, M., Ng, A.M. and Shridharan P. (2012) *International Journal of Pharmaceutics*, **430**(1–2): 299–306.

Tiwari, A, Tiwari, A. and Singh, R.P. (2012) In: Tiwari, A., Ramalingam, M., Kobayashi, H. and Turner, A.P.F. (Eds) *Biomedical Materials and Diagnostic Devices*, Scrivener-Wiley, Beverly, pp. 303–322.

Torchilin, V.P. (2001) *Biochimica et Biophysica Acta*, **1511**(2): 397–411.

Üner, M. and Yener, G. (2007) *International Journal of Nanomedicine*, **2**(3): 289–300.

Valtcheva-Sarker, R.V., O'Reilly, J.D. and Sarker, D.K. (2007) *Recent Patents on Drug Delivery and Formulation*, **1**(2): 147–159.

van der Valk, A., Howbrook, D., O'Shaughnessy, M., Sarker, D., Baker, S.C., Louwrier, A. and Lloyd, A. (2003) *Biotechnology Letters*, **25**(16): 1325–1328.

Vauthier, C. and Couvrer, P. (2007) *PharmTech Europe*, **19**(1): 35–42.

Videira, M.A., Botelho, M.F., Santos, A.C., Gouveia, L.F., de Lima, J.J. and Almeida, A.J. (2002) *Journal of Drug Targeting*, **10**: 607–613.

Waghmaree, A., Deopurkar, R.L., Salvi, N., Khadilkar, M., Kalolikar, M. and Gade, S.K. (2009) *Vaccine*, **27**(7): 1067–1072.

Wang, J.J., Sung, K.C., Hu, O.Y., Yeh, C.H. and Fang, J.Y. (2006) *Journal of Controlled Release*, **115**(2): 140–149.

Wang, J.J., Hung, C.F., Yeh, C.H. and Fang, J.Y. (2008) *Journal of Drug Targeting*, **16**(4): 294–301.

Wang, Q., Tan, G., Lawson, L.B., John, V.T. and Papadopoulos, K.D. (2010) *Langmuir*, **26**(5): 3225–3231.

Wiessleder, R., Heautot, J.F., Schaffer, B.K., Nossiff, N., Papisov, M.I., Bogdanov, A.A. and Brady, T.J. (1994) *Radiology*, **191**: 225–230.

Wilde, P.J. and Clark, D.C. (1993) *Journal of Colloid and Interface Science*, **155**: 48–54.

Woodward N.C., Gunning A.P., Mackie A.R., Wilde P.J. and Morris V.J. (2009) *Langmuir*, **25** (12): 6739–6744.

Wüstneck, R., Krägel, J., Miller, R., Fainerman, V.B., Wilde, P.J., Sarker, D. and Clark, D.C. (1996) *Food Hydrocolloids*, **10**(4): 395–405.

Yguerabide, J., Schmidt, J.A. and Yguerabide, J. (1982) *Biophysics Journal*, **79**: 69–75.

Young, D.L. and Michelson, S. (2012) *Systems Biology in Drug Discovery and Development*, John Wiley & Sons, Hoboken.

Index

Pharmaceutical Emulsions: A Drug Developer's Toolbag, First Edition. Dipak K. Sarker.
© 2013 John Wiley & Sons, Ltd. Published 2013 by John Wiley & Sons, Ltd.